SpringerBriefs in Electrical and Computer Engineering

For further volumes:
http://www.springer.com/series/10059

Jukka Suhonen • Mikko Kohvakka • Ville Kaseva
Timo D. Hämäläinen • Marko Hännikäinen

Low-Power Wireless Sensor Networks

Protocols, Services and Applications

Jukka Suhonen
Tampere University of Technology
Finland
jukka.suhonen@tut.fi

Mikko Kohvakka
Suntrica Ltd
Tampere
Finland

Ville Kaseva
Tampere University of Technology
Finland

Timo D. Hämäläinen
Tampere University of Technology
Finland

Marko Hännikäinen
Tampere University of Technology
Finland

ISSN 2191-8112   e-ISSN 2191-8120
ISBN 978-1-4614-2172-6   e-ISBN 978-1-4614-2173-3
DOI 10.1007/978-1-4614-2173-3
Springer New York Dordrecht Heidelberg London

Library of Congress Control Number: 2011946225

Printed on acid-free paper

Springer is part of Springer Science+Business Media (www.springer.com)

Preface

Wireless sensor networks (WSN) is a family of technologies which combine sensing, data processing, and wireless ad-hoc networking, and enable new application scenarios. A networked sensor can be a radar station, pollution monitor, video camera, mobile phone, heart rate monitor, or an ID tag.

In this book, we take resource contained WSN technologies into study. These kinds of WSN nodes are small, cheap, operate on batteries, and automatically form networks of thousands of measurements points. Together, the nodes form a distributed platform performing monitoring, object tracking, and control functions. These kinds of ubiquitous WSNs are an enabling technology for environmental and condition monitoring, home automation, security and alarm systems, industrial monitoring and control, military reconnaissance and targeting, and interactive games, to mention just a few.

This book describes low-power WSN as a platform by presenting the WSN services that can be used as building blocks for the applications. It explains the implications of resource constraints and expected performance in terms of throughput, reliability and latency.

The book builds on our experiences on developing WSN platforms, protocols, and prototyping with different applications. The book eases going through the vast design space of WSNs when putting together the platforms, communications, and application requirements. The book is a concise report of the state of the art in resource contained WSNs, making it easier for students, engineers, and researchers to adopt this emerging area of technology.

Tampere, Finland, *Authors*
November 2011

Contents

Acronyms

6LoWPAN	IPv6 over Low power Wireless Personal Area Networks
ACL	Access Control List
ADC	Analog-to-Digital Converter
API	Application Programming Interface
APS	Application Support
ASIC	Application Specific Integrated Circuit
BAN	Body Area Network
CAN	Controller Area Network
CAP	Contention Access Period
CDMA	Code Division Multiple Access
CFP	Contention-Free Period
COTS	Commercial Off-The-Shelf
CRC	Cyclic Redundancy Check
CSMA/CA	Carrier Sense Multiple Access with Collision Avoidance
CSS	Chirp Spread Spectrum
CTS	Clear-To-Send
DC	Direct Current
DMTS	Delay Measurement Time Synchronization
DSN	Deployment Support Network
EEPROM	Electrically Erasable Programmable Read-Only Memory
ERP	Election/Re-election Procedure
FDMA	Frequency Division Multiple Access
FFD	Full Function Device
GPS	Global Positioning System
GPRS	General Packet Radio Service
GSM	Global System for Mobile Communications
HID	Human Interface Device
HTTP	Hypertext Transfer Protocol
HRTS	Hierarchy Referencing Time Synchronization
IETF	Internet Engineering Task Force
IP	Internet Protocol

IPC	Inter-Process Communication
ITR	Individual-based Time Request
LAN	Local Area Network
LED	Light Emitting Diode
LOS	Line-of-Sight
LPL	Low Power Listening
LR-WPAN	Low-Rate Wireless Personal Area Network
LQI	Link Quality Indication
LTS	Lightweight Tree-based Synchronization
MAC	Medium Access Control
MCU	Micro-Controller Unit
MIPS	Million Instructions Per Second
NTP	Network Time Protocol
OGC	Open Geospatial Consortium
OS	Operating System
PAN	Personal Area Network
PHY	Physical
QoS	Quality of Service
RBS	Reference Broadcast Synchronization
RDP	Rate-based Diffusion Protocol
RPC	Remote Procedure Call
RF	Radio Frequency
RFD	Reduced Function Device
RFID	Radio Frequency Identification
RPC	Remote Procedure Call
RSS	Received Signal Strength
RSSI	Received Signal Strength Indicator
RTS	Request To Send
RTT	Round Trip Time
SKKE	Symmetric-Key Key Exchange
SoC	System-on-Chip
SQL	Structured Query Language
SRAM	Static Random Access Memory
SKKE	Symmetric-Key Key Exchange
SWE	Sensor Web Enablement
TDMA	Time Division Multiple Access
TDP	Time-Diffusion synchronization Protocol
TEDS	Transducer Electronic Data Sheet
TIM	Transducer Interface Module
TPSN	Timing-sync Protocol for Sensor Networks
UMTS	Universal Mobile Telecommunications System
UDP	User Datagram Protocol
UTC	Universal Time Coordinated
UWB	Ultra Wide Band
VLC	Visible Light Communications

WLAN	Wireless Local Area Network
WMAN	Wireless Metropolitan Area Network
WPAN	Wireless Personal Area Network
WSN	Wireless Sensor Network
WWAN	Wireless Wide Area Network
XML	eXtensible Markup Language

Chapter 1
Low-power WSN Technology

Generally, a sensor node refers to any device that is capable to sense its environment. Wireless Sensor Network (WSN) as a technology is a collection of sensor devices that co-operate with each other. A WSN may comprise even thousands of autonomic and self-organizing nodes that combine environmental sensing, data processing, and wireless multihop ad-hoc networking. The features of WSNs enable monitoring, object tracking, and control functionality. The potential applications include environmental and condition monitoring, home automation, security and alarm systems, industrial monitoring and control, military reconnaissance and targeting, and interactive games.

Low-power WSNs are characterized by extremely low cost and ultra low energy [9]. This allows the deployment of potentially disposable devices that can have a battery powered lifetime of years or operate on energy gathered from their environment. However, as a trade-off, the low-power WSNs have limited computation, communication, memory, and energy resources.

1.1 WSN and Other Wireless Technologies

Wireless communication technologies are categorized based on their typical coverage and application domains. The link range, data rate, mobility, and power requirements of the technologies are presented in Fig. 1.1. The values are not definite but illustrate the differences between the technologies. In the figure, Radio Frequency (RF) communications is assumed as it is most widely used and does not have inherent limitations such as line-of-sight requirement in infrared.

Wireless Wide Area Network (WWAN) covers a large geographical area and consists of telecommunications networks such as Global System for Mobile Communications (GSM) and satellite communications. In telephone networks, broadband data is supported with packet-switched data services such as General Packet Radio Service (GPRS), 3G, or Universal Mobile Telecommunications System (UMTS).

Mobility requirements are critical, as uninterrupted service is expected even when a user is traveling on high-speed rail (200+ km/h).

Wireless Metropolitan Area Network (WMAN) covers geographic area or region that is smaller than WWAN but larger than Wireless Local Area Network (WLAN). An example of WMANs is IEEE 802.16 (WiMAX) [6]. Both WWAN and WMAN use highly asymmetric devices, as simpler end devices connect to base stations. As such, these networks are intended for single hop uses where the wireless access is used to connect to the Internet or global telephone network. Wireless multihop support is rare and typically limited to base stations.

WLAN spans a relatively small area, such as building or a group of buildings. IEEE 802.11 [3] is the dominant WLAN technology. It was originally targeted to access a wired Local Area Network (LAN) with wireless interface but has been since extended to support mesh networking. IEEE 802.11 is widely utilized for network access in public buildings and enterprises, and sharing Internet in homes.

Wireless Personal Area Network (WPAN) is a short distance network for interconnecting devices centered around an individual person including watches, headsets, mobile phones, audio/video equipment, and laptops. Bluetooth [2] and IEEE 802.15 standard family [4, 5] are the most widely used WPAN technologies. WPANs have varying energy and throughput requirements as the use cases range from low power data exchange with portable devices to high data rate home entertainment and multimedia transfers.

WSN shares most properties with WPANs and may utilize similar technologies. For example, IEEE 802.15.4 low-rate WPAN standard [5] is used as a basis for many WSN communication standards. However, a WSN is designed for multiple users, has usually more devices, and often emphasizes lifetime.

1.2 Characteristics of Low-power WSNs

A WSN consists of nodes that are deployed in the vicinity of an inspected phenomenon [1] as shown in Fig. 1.2. In addition, a network may contain one or more

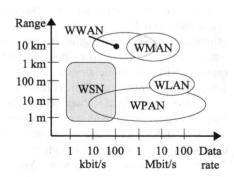

Fig. 1.1 Properties of wireless communication technologies.

sink nodes that request other nodes to perform measurements and collect the measured values for further use. Instead of sending raw data to the sink, a sensor node may collaborate with its neighbors or nodes along the routing path to provide application results [8]. The sink node typically acts as a gateway to other networks and user interfaces [7]. The backbone infrastructure that is connected to a sink may contain components for data storing, visualization, and network control.

Compared to the traditional computer networks, WSNs have several unique characteristics as listed in the following. While a WSN may not share every characteristic, e.g. the scalability is not a primary concern on few nodes deployments, many useful classes of WSNs share most or all of the properties described in this list.

- *Network size and density*: WSNs may consist of tens of thousands of nodes. The density of nodes can be high, depending on the application requirements for sensing coverage and robustness via redundancy.
- *Communication paradigm*: In WSNs, node identifiers are typically not important. Instead WSNs are *data-centric*, which means that messages are not send to individual nodes but to geographical locations or regions based on the data content.
- *Application specific*: A WSN is deployed to perform a specific task, e.g. environmental monitoring, target tracking, or intruder alerting. As a result, the node platforms and communication protocols are designed to optimal performance on a certain application-dependent scenario. The application specific behavior enables data aggregation, and in-network processing, and decision making.
- *Network lifetime*: WSNs are typically deployed to observe certain physical phenomenon that range in duration from fractions of a second to a few months or even several years. As replacing batteries is not feasible due to large network size and deployment to possibly hazardous environment, nodes must optimize their energy usage for network lifetime.

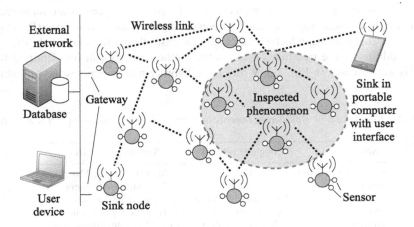

Fig. 1.2 A typical WSN scenario.

- *Low cost*: To allow cost effective deployment of a large number of nodes, the cost of an individual sensor node should be minimized. Also, as recovering sensors after deployment in some application scenarios may not be feasible, sensors should be cheap enough to be considered disposable.
- *Resource constraints*: A typical WSN node combines low cost with small physical size and is battery powered. Thus, computation, communication, memory, and energy resources are very limited.
- *Dynamic nature*: Wireless communications are inherently unreliable due to environmental interferences. The unreliability is especially evident in WSNs because of harsh operating conditions e.g. due to environmental changes in outdoors, node mobility, and nodes dying due to depleted energy sources. As a result, the unreliability causes network dynamics due to link breaks even when nodes are stationary.
- *Deployment*: To avoid tedious network planning of a large number of nodes, WSNs are often randomly deployed. This necessitates network self-configuration and autonomous operation.

1.3 Quality of Service Requirements

Quality of Service (QoS) is commonly expressed and managed by throughput, latency, jitter, and reliability. These QoS parameters also apply to the WSNs, but their importance differs from the legacy networks.

The requirements of low-energy WSNs compared to the traditional wireless computer networks, e.g. IEEE 802.11 Wireless LAN (WLAN), are summarized in Table 1.1. Some sensing applications can tolerate high latency and low throughput but the reliability is particularly significant. In the traditional computer networks, the data is routed via highly reliable wired links, while only the end links may be wireless, utilizing e.g. cellular connections or WLAN. In WSNs, packets are forwarded via multiple wireless hops. On each wireless link, the packet error rates (PER) of 10%-30% are common, which significantly decreases the end-to-end reliability.

In addition to the traditional QoS metrics, other metrics can be identified for WSNs as presented in Fig. 1.3. While the reliability metric denotes the probabil-

Table 1.1 Comparison of typical requirements in wireless computer networks and low-energy WSNs.

Requirement	Computer network	Low-energy WSN
Resource constraints	Low	Very high (1-2 MIPS, 32-128 kB)
Adaptivity	Static	Dynamic environment
Scalability	Moderate (10 nodes)	High (10000 nodes)
Latency	High (250 ms-1 s)	High-Low (1 s-1 hour)
Throughput	Very high (MB/s)	Low-Moderate (bit/s-kbit/s)

ity to transfer a single measurement through the network, the availability expresses the probability to receive a new measurement from a node within a certain waiting period [10]. The data accuracy describes the consistency of measurements (results are same in similar conditions) and the granularity of sensor values, sensing location, and time information. The security ensures that unauthorized parties do not gain access or tamper with the sensed data. The mobility is important in tracking WSNs as a node may be attached to moving objects. Due to the significance of the network lifetime, energy efficiency is considered as a QoS parameter. Usually, the other parameters have a trade-off with the energy efficiency, making it impossible to optimize all parameters at the same time.

As an example, the QoS requirements for two cases are presented in Fig. 1.3(a) and Fig. 1.3. An environmental monitoring application sends sensor values periodically to a sink. As the sensed environment changes slowly, the measurement interval and thus throughput can be low, while missing a single sample is not critical. However, availability is still important, as too many samples may not be missed consecutively. In a control network, short messages are relayed infrequently between switches, lamps, and other accessories. Thus, the required bandwidth is low, but the timely and reliable delivery of commands is required.

Due to the diversity of applications and their contradictory requirements, a single solution is not suitable for every WSN application. Thus, the protocols and node platforms need to be tailored to meet the application requirements.

1.4 Services for Sensing and Actuation Applications

A WSN offers following services for sensing and actuation applications.

- *Environmental sensing*: Each WSN node contains at least one or several physical sensors. Instead of accessing nodes directly, e.g. via Inter-Integrated Circuit (I2C) bus, sensing services operate via standardized interfaces.

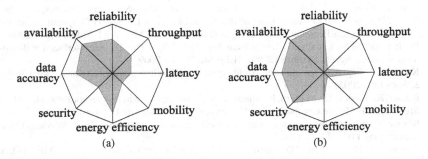

Fig. 1.3 Quality of Service (QoS) parameters in WSNs. **a** Typical environmental monitoring network emphasizing energy efficiency. **b** A control network emphasizing low latency and reliability.

- *Data processing and storage*: Before the sensor values can be forwarded to a user, the values are preprocessed and stored locally. The limitations of computing and storage can be overcome by distributed computing services.
- *Data transfer*: Data transfer services allow collecting data for further use. The realized QoS is largely affected by the choice of networking protocols.
- *Localization*: A sensor value is naturally associated with a certain area. However, random deployment and mobility prevent concluding the location information from source node identifiers, which necessitates either online (distributed) or offline (centralized) localization services.
- *Time synchronization*: Several WSN applications, such as event alerts and tracking, require exact timestamping of sensor events to compare the order of events. As low cost sensor nodes do not have accurate time source, an agreement on global time is achieved with time synchronization service.

A WSN provides at least the sensing, data processing, and data transfer services. The localization and time synchronization services are not usually considered in proposed WSN standards but need customized solutions.

References

1. Akyildiz, I.F., Su, W., Sankarasubramaniam, Y., Cayirci, E.: Wireless sensor networks: a survey. Elsevier Computer Networks **38**(4), 393–422 (2002)
2. Bluetooth Special Interest Group (SIG) Std. 2.0 + EDR: Specification of the Bluetooth System (2004)
3. Information Technology—Telecommunications and information exchange between systems—Local and metropolitan area networks—Specific requirements—Part 11: Wireless LAN Medium Access Control (MAC) and Physical Layer (PHY) specifications (1997). IEEE Std 802.11-1997
4. IEEE Standard for Information Technology—Telecommunications and Information Exchange Between Systems—Local and Metropolitan Area Networks—Specific Requirements—Part 15.3: Wireless Medium Access Control (MAC) and Physical Layer (PHY) Specifications for High Rate Wireless Personal Area Networks (WPANs) (2003). IEEE Std 802.15.4-2003
5. IEEE Standard for Information Technology—Telecommunications and Information Exchange Between Systems—Local and Metropolitan Area Networks—Specific Requirements—Part 15.4: Wireless Medium Access Control (MAC) and Physical Layer (PHY) Specifications for Low-Rate Wireless Personal Area Networks (LR-WPAN) (2006). IEEE Std 802.15.4-2006
6. IEEE Standard for Local and Metropolitan Area Networks—Part 16: Air Interface for Broadband Wireless Access Systems (2009). IEEE Std 802.16-2009
7. Karl, H., Willig, A.: Protocols and Architectures for Wireless Sensor Networks. John Wiley & Sons Ltd (2005)
8. Römer, K., Kasten, O., Mattern, F.: Middleware challenges for wireless sensor networks. ACM SIGMOBILE Mobile Computing and Communications Review **6**(4), 59–61 (2002)
9. Römer, K., Mattern, F.: The design space of wireless sensor networks. IEEE Wireless Communications **11**(6), 54–61 (2004)
10. Suhonen, J., Hämäläinen, T.D., Hännikäinen, M.: Availability and end-to-end reliability in low duty cycle multihop wireless sensor networks. Sensors **9**(3), 2088–2116 (2009)

Chapter 2
Key Standards and Industry Specifications

A vast number of sensors exists in the sensor industry that can be used to measure physical parameters, such as temperature, pressure, humidity, illumination, gas, flow rate, strain, and acidity. Standardized sensor interfaces, data formats, and communication protocols are required to enable effective integration, access, fusion, and the use of sensor-derived data. The goal is to allow sensors from different manufacturers to work together without human intervention and customization.

Standards have several advantages and safeguards compared to proprietary technologies. The technology has the support from industry, thus bolstering customer confidence and allowing larger markets. Also, a standard allows future proofing as the technology is not dependent on a single vendor. The public availability of specifications draws research from multiple sources. This ensures that the standard technologies are continually improving in their performance, and expanding into new areas and applications.

Existing wireless standards, such as IEEE 802.11 [4] WLAN or GSM standards, can be used in conjunction with the sensor networks. However, these are not designed for WSNs and do not take account for their unique characteristics. Thus, this section considers only the communication standards that are specifically targeted for low power WSNs.

The operating frequency band, nominal data rate, and protocol support of WSN communication standards are listed in Table 2.1. The support for PHYsical (PHY), Medium Access Control (MAC), Network (NWK), and Transport (TRP) protocols denotes that a standard defines the layer in question. Application Support (APS) defines application profiles detailing the services, message formats, and methods required to access applications, therefore allowing interoperability between devices from different manufacturers. For security, Access Control Lists (ACLs)allow only certain nodes to participate in the network while data encryption prevents unauthorized use of data. The listed standards use 128-bit Advanced Encryption System (AES) for data encryption.

Table 2.1 The properties of WSN communication standards.

Standard	Frequency (MHz)	Data rate (kbps)	PHY	MAC	NWK	TRP	APS	ACL	Encryption
IEEE 802.15.4	868	20	●	●	○	○	○	●	●
	915	40	●	●	○	○	○	●	●
	2400	250	●	●	○	○	○	●	●
ZigBee[1]	-	-	○	○	●	●	●	●	●
6LoWPAN[1]	-	-	○	○	●	○	○	●	●
WirelessHART[2]	2400	250	●	●	●	●	●	●	●
ISA100.11a[2]	2400	250	●	●	●	●	●	●	●
Z-Wave	865	40	●	●	●	○	●	○	○
	915	40	●	●	●	○	●	○	○
Bluetooth Low Energy	2400	1000	●	●	●	●	●	●	●
ANT/ANT+	2400	1000	●	●	●	○	●	●	○
ONE NET	868/	38.4	○	●	●	○	○	●	●
	915	230	○	●	●	○	○	●	●
DASH7	433	27.8	●	●	○	○	○	○	○
IEEE 1902.1 RuBee	0.131	1.2	●	●	○	○	○	○	○

[1] Uses IEEE 802.15.4 as PHY and MAC layers
[2] Uses IEEE 802.15.4 as PHY layer

2.1 IEEE 802.15 Standard Family

IEEE 802.15 family defines several standards targeted at WPANs:

- *802.15.1*: lower layers of Bluetooth 1.x protocol stack
- *802.15.2*: methods to improve coexistence of WPANs with other wireless devices in unlicensed frequency bands. The proposed methods include adaptive frequency hopping, power control, and techniques to avoid same frequencies
- *802.15.3*: high speed WPAN for multimedia applications with data rates from 55 MB/s up to several GB/s
- *802.15.4*: Physical (PHY) and Medium Access Control (MAC) layers for Low-Rate Wireless Personal Area Networks (LR-WPANs)
- *802.15.5*: mesh networking functionality for WPANs
- *802.15.6*: Body Area Network (BAN) which operate in and around the human body
- *802.15.7*: PHY/MAC for Visible Light Communications (VLC)

In the standard family, IEEE 802.15.4 [5] and IEEE 802.15.5 standards are the most relevant for the resource constrained WSNs.

IEEE 802.15.4 is relatively popular standard and has therefore been used as a basis for other standards. Fig. 2.1 presents how different standards build on top of the IEEE 802.15.4. While few standards, such as IEEE 802.15.5 and ZigBee[13], use the IEEE 802.15.4 as a whole, other presented standards reuse the PHY layer. The motion is to save money by using compatible transceivers that are already popular and on the market.

IEEE 802.15.4 network supports three types of network devices: a Personal Area Network (PAN) coordinator, coordinators, and devices. The PAN coordinator initiates the network and operates often as a gateway to other networks. Coordinators collaborate with each other for data routing and network self-organization. Devices do not have data routing capability and can communicate only with coordinators.

IEEE 802.15.4 defines two types of devices, a Full Function Device (FFD) that can act as a router and a Reduced Function Device (RFD) that can communicate only with a FFD. Thus, a RFD is intended for simple use, e.g. to act as an light switch, and cheaper to implement as it requires less memory and processing capability. A 802.15.4 network always has one FFD device that act as a PAN coordinator. The PAN coordinator initiates the network and operates often as a gateway to other networks. The relation between FFDs and RFDs, and the supported star and peer-to-peer (mesh) topologies are show in Fig. 2.2.

IEEE 802.15.4a extension to the standard defines two optional high-rate PHYs, Ultra Wide Band (UWB) PHY operating in the sub-gigahertz band and in 3 GHz to 10 GHz frequency band and Chirp Spread Spectrum (CSS) operating in the 2.4 GHz band. Boths PHYs enhance robustness via resistance to multipath fading while operating with low power (0 dBm). The UWB PHY has the nominal over-the-air data rate of 851 kbit/s, and the optional data rates of 110 kbit/s, 6.81 Mbit/s, and 27.24 Mbit/s. Utilizing the unique capabilities of UWB waveforms, the UWB PHY can be used for precision ranging (1 m accuracy) between devices. The CSS PHY has the nominal data rate of 1000 kbit/s and optionally can support 250 kbit/s data

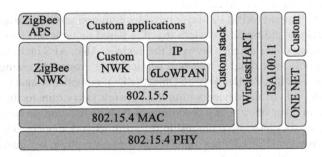

Fig. 2.1 IEEE 802.15.4 related protocols.

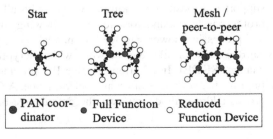

Fig. 2.2 Network topologies in IEEE 802.15.4.

rate. The CSS PHY has long link range allowing the support for mobile devices moving at higher speeds.

IEEE 802.15.5 defines two mesh network frameworks, a high-rate framework for IEEE 802.15.3 and a low-rate framework for IEEE 802.15.4. Both frameworks define procedures for topology formation and data routing. In addition, the low-rate framework describes mobility support (e.g. route maintenance), the practical use of power saving modes of IEEE 802.15.4, and a trace route service to monitor the status of a route.

2.2 ZigBee

ZigBee standard defines network and application layers on top of the IEEE 802.15.4 [13]. These layers comprise of several sub-layers as shown in Fig. 2.3. ZigBee defines a wide range of application profiles targeted at home and building automation, remote controls, and health care.

The ZigBee network layer connects application objects with end points that allow distinguishing between applications, and are akin to the TCP/UDP ports in IP networking. A network can use star, peer-to-peer, or cluster-tree topologies. A network has always one device referred to as a ZigBee coordinator that controls the network. The coordinator is the central node in the star topology, the root of the tree in the tree topology, and can be located anywhere in the peer-to-peer topology. To reduce interference, ZigBee uses a network wide frequency agility mechanism where an individual device can report the interference and instruct network coordinator to change the network wide used channel.

Application Support (APS) sub-layer provides common services for applications. It allows matching and binding devices together based on their services and needs, this way allowing data exchange between the devices. Also, APS includes fragmentation, reassembly, and reliable transport services and group addressing to enable multicast functionality. ZigBee Device Object (ZDO) component manages network by defining the role of the device within the network (a coordinator or an end device), discovers other devices and determines which application services they provide, and establishes a secure relationship between network devices. ZigBee Security Service Provider (SSP) defines methods for data encryption, key generation, key distribution, and authentication.

ZigBee defines two stack profiles, ZigBee and ZigBee Pro. Both stack profiles support full mesh networking and work with all application profiles. The ZigBee profile has less mandatory features, allowing implementation with less memory and processing power. ZigBee Pro mandates many features that are optional in the ZigBee stack profile, e.g. network wide encryption via Symmetric-Key Key Exchange (SKKE). In addition, ZigBee Pro provides additional features that aim to improve network scalability and performance. A stochastic address assignment improves address allocation over the default tree-based assignment, thus improving network scalability. Many-to-one routing support improves routing efficiency when a network has only few data collecting sinks.

2.3 6LoWPAN

IPv6 over Low power Wireless Personal Area Networks (6LoWPAN) [7] defines the transmission of IP packets over IEEE 802.15.4 networks. This way, the sensor network can act seamlessly as a part of the global Internet, thus enabling the use of wide range of existing IP-based applications and technologies. The standard does not define any routing protocols but relies on other specifications, such as e.g. IEEE 802.15.5, for mesh layer routing. 6LoWPAN is described in detail in Chapter 6.

2.4 Z-Wave

Z-Wave [12] is targeted at building automation and entertainment electronics and has been developed by over 120 companies including Zensys, Intel and Cisco. A Z-Wave network comprises controlling devices (e.g. remote controller) and AC powered slave nodes (e.g. television). Slave nodes can act as routers. As routers continuously listen to the wireless medium for incoming transmission, they need to be mains powered in practice. Controllers are active on demand and can be battery powered. The protocol operation is simple, a controller hosts a routing table to the entire network. If a packet needs to be routed over multiple hops, the controller embeds the route to the forwarded packet. The maximum number of nodes in a network is 232, although Z-wave networks can be inter-connected via gateways.

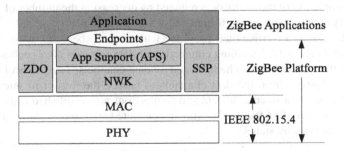

Fig. 2.3 ZigBee defines network and application layers on top of IEEE 802.15.4 standard.

2.5 WirelessHART and ISA100.11a

WirelessHART and ISA100.11a [6] are targeted at process industry applications where process measurement and control applications have stringent requirements for end-to-end communication delay, reliability, and security. Although the exact communication formats are different, the standards have similar operating principle. The convergence of the standards is planned in ISA100.12.

Both standards build on top of the IEEE 802.15.4 physical layer and utilize a TDMA MAC that employs network wide time synchronization, channel hopping, channel blacklisting. The Time Division Multiple Access (TDMA) based operation allows accurately determining bounds for latency and throughput, therefore being predictability which is important for industrial applications. A centralized network manager is responsible for route updates and communication scheduling for entire network. However, the centralized control of TDMA schedules limits the network size and the tolerance against network dynamics, thus limiting the usability of the standards to static WSNs.

2.6 Bluetooth Low Energy

Bluetooth Low Energy (BLE) [1] is an extension to the Bluetooth technology aimed at low energy wireless devices. The first defined applications comprise watch, Human Interface Device (HID), and sensor profiles. Despite the name, BLE is not compatible with the traditional Bluetooth due to different link layer protocol. Compared to the traditional Bluetooth, the main functional differences are the use of variable packet length, entering power save mode automatically when a device is not transmitting, and the exchange data in attribute/value pairs. These changes aim at minimizing transceiver's active time. To ease resource constrained implementation, the complexity of the protocol is reduced by decreasing the number of different connectivity states and message formats.

In BLE, devices advertise their presence with periodic beacons, while listening to the channel briefly for incoming connection or data requests after each advertisement. BLE uses 3 advertising channels and 37 data channels. The protocol employs frequency hopping within the data channels to reduce the impact of interference. Advertisements can also contain data and connections are established fast (less than 3 ms), therefore avoiding the need to stay in connected state and enabling devices to save energy in standby states.

2.7 ANT and ANT+

ANT [2] defined by Dynastream Innovations Inc. is used e.g. by Suunto and Garmin in their performance monitoring products. Its communication model is based on virtual channels that are defined by operating frequency and message rate parameters. The medium is accessed with TDMA where time slots are repeated on configured message intervals. As a result, several channels may operate on the same physical frequency.

Each channel has a single master device and slave devices. The channel itself can be unidirectional from a master to a slave, unidirectional from a slave to a master, or bidirectional. Further, the channel can be configured as a broadcast channel, use optional acknowledgments, or have burst support. As each node may act both as a master and a slave on different channels, ANT allows forming very complex topologies. However, considering that to build optimal TDMA schedules, traffic characteristics must be know a priori. Thus, ANT is best suited for relatively small networks.

ANT+ is an extension to the ANT protocol that uses the same communication method but includes profiles defining data formats and channel parameters.

2.8 ONE-NET

ONE-NET is an open source WSN specification that uses IEEE 802.15.4 compatible transceivers while specifying MAC and routing layers [11]. The protocol operates with the basic data rate of 38.4 kbit/s, although the specification allows up to 230 kbit/s data rates. In ONE-NET, all transmissions are encrypted with XTEA2 algorithm. Other security features include user key management. ONE-NET supports low duty cycling for battery powered devices but routing nodes must keep their transceivers active thus necessitating mains power. ONE-NET has limited APS with generic message formats for applications, but these do not define formats for specific applications to guarantee compatibility between implementations.

2.9 DASH7

DASH7 [10] technology based on ISO 18000-7 standard is targeted at very low rate data applications. Its main cited benefit stems from the 433 MHz operating frequency, which provides longer communication ranges and less crowded wireless channel than the typical 2.4 GHz frequency band [8]. DASH7 has the nominal communication range of 250 m at 0 dBm transmission power level, compared to 75 m of ZigBee and 10 m of Bluetooth (High Rate variant) [8].

2.10 IEEE 1902.1

IEEE 1902.1 (RuBee) [3] [9] fills the gap between WSN and Radio Frequency Identification (RFID) technologies. Unlike other listed technologies, signal does not include electric field component but uses magnetic dipole antennas. Thus, signal is unaffected by water and metals either enhance or do not affect the signal. RuBee nodes, referred to as tags, can be very simple identity tags or use 4-bit MCU, 0.5 kB-2 kB Static Random Access Memory (SRAM), optional sensors, signal processing firmware, displays and buttons [9]. The nominal data rate is small, 1.2 kbps, limiting the applicability of RuBee.

2.11 IEEE 1451

IEEE 1451 standard family defines a set of open, network-independent communication interfaces for connecting transducers (sensors and actuators) to microprocessors, instrumentation systems and networks. In IEEE 1451, a single sensor, an actuator, or a module comprising several transducers and any data conversion or signal conditioning (e.g. signal amplification or filtering) is referred to as a Transducer Interface Module (TIM). As a key element, the standard family defines Transducer Electronic Data Sheet (TEDS) to store transducer identification, calibration, correction data, and manufacturer related information. It also defines a set of commands to control and read data from the TIM. In practice, TEDS can be a software module integrated within a TIM. TEDS practically eliminates error prone, manual entering of data and system configuration, and allows transducers to be installed, upgraded, replaced or moved by plug-and-play principle.

IEEE 1451.0 and IEEE 1451.1 standards define generic methods to access transducers, and define transducer related interfaces and services. In addition, IEEE 1451 provides interfaces for several standardized communication protocols in IEEE 1451.2-1451.6 standards. IEEE 1451.2 defines wired point-to-point communication. IEEE 1451.3 defines distributed multi-drop system, where a large number of TIMs may be connected along a wired multi-drop bus. IEEE 1451.4 specifies mixed-mode communication protocols, which carry analog sensor values with digital TEDS data. IEEE 1451.6 defines a high-speed Controller Area Network (CAN) bus.

IEEE 1451.5 standard defines wireless sensors and thus, it is most closely related with WSNs. Supported communication technologies are IEEE 802.11a/b/g, IEEE 801.15.1, and IEEE 802.15.4.

References

1. Bluetooth SIG: Bluetooth Specification Version 4.0 (2009)

2. Dynastream Innovations Inc.: ANT Message Protocol and Usage Rev 3.1 (2009). URL `http://thisisant.com`. D00000652
3. IEEE Standard for Long Wavelength Wireless Network Protocol (2009). DOI 10.1109/IEEESTD.2009.4810102. IEEE Std 1902.1-2009
4. Information Technology—Telecommunications and information exchange between systems—Local and metropolitan area networks—Specific requirements—Part 11: Wireless LAN Medium Access Control (MAC) and Physical Layer (PHY) specifications (1997). IEEE Std 802.11-1997
5. IEEE Standard for Information Technology—Telecommunications and Information Exchange Between Systems—Local and Metropolitan Area Networks—Specific Requirements—Part 15.4: Wireless Medium Access Control (MAC) and Physical Layer (PHY) Specifications for Low-Rate Wireless Personal Area Networks (LR-WPAN) (2006). IEEE Std 802.15.4-2006
6. ISA: ISA100.11a release 1. Available: `http://www.isa.org/source/ISA100.11a_Release1_Status.ppt` (2007)
7. Montenegro, G., Kushalnagar, N., Hui, J., Culler, D.: Transmission of IPv6 packets over IEEE 802.15.4 networks. RFC 4944 (2007)
8. Norair, J.P.: Introduction to DASH7 technologies. Tech. rep., DASH7 Technology Working Group (2009)
9. An introduction to RuBee technology (2010). URL `http://www.rubee.com/Partners/Oracle/RuBeeWhitePaper-v3.pdf`
10. Schneider, D.: Wireless networking dashes in a new direction. IEEE Spectrum **47**(2), 9–10 (2010). DOI 10.1109/MSPEC.2010.5397768
11. Threshold Corporation: ONE-NET Specification (2009). Version 1.5.0
12. Zensys A/S: Z-Wave Protocol Overview (2007). URL `http://www.zen-sys.com/`. Doc. No. SDS10243-4
13. ZigBee Standards Organization: ZigBee Specification (2008). ZigBee Document 053474r17

Chapter 3
Hardware Platforms and Components

Sensor node platforms implement the physical layer (hardware) of the protocol stack. The hardware activity measured as the fraction of time the hardware is in an active state (processing data or receiving/transmitting a packet) may be below 1% in low data-rate monitoring applications. Thus, it is very important to minimize the power consumption in idle and sleep modes.

A general hardware architecture of a sensor node platform is presented in Fig. 3.1. The architecture can be divided into four subsystems:

- *Communication subsystem* enabling wireless communication,
- *Computing subsystem* allowing data processing and the management of node functionality,
- *Sensing subsystem* connecting the wireless sensor node to the outside world, and
- *Power subsystem* providing the system supply voltage.

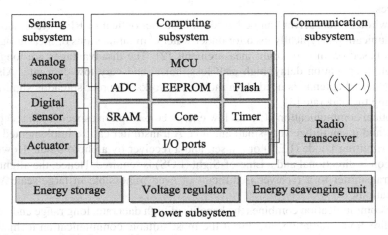

Fig. 3.1 Sensor node hardware architecture.

17

This chapter focuses on Commercial Off-The-Shelf (COTS) hardware components and their typical features and performance. When designing own platforms, readily available COTS components can significantly reduce development time when compared e.g. to Application Specific Integrated Circuit (ASIC) designs. In addition, some of the commercially available standard compliant transceivers implement part of the protocol stack, thus significantly reducing the development effort.

3.1 Communication subsystem

The communication subsystem consists of a wireless transceiver and an antenna that are used to transmit and receive messages one bit or symbol at a time. The functions available in most transceivers are the selection of a frequency channel and a transmit power, the modulation transmitted and demodulation of received data, symbol synchronization and clock generation for received data.

A transceiver may also include additional functions, which reduce the processing requirements of Micro-Controller Unit (MCU). For example, an IEEE 802.15.4 compliant PHY includes data frame synchronization for perceiving the start of an incoming frame, clear channel assessment for detecting ongoing traffic in a frequency channel, Received Signal Strength Indicator (RSSI) and Link Quality Indication (LQI) for measuring signal strength and estimating link quality to neighboring nodes, Cyclic Redundancy Check (CRC) calculation for checking bit errors on received frames, data encryption/decryption for improving network security and automatic acknowledge transmissions after received frames. Since these features are implemented most efficiently in physical layer, they can improve overall network energy-efficiently. Yet, the increased complexity raises hardware cost. In practice, the lowest power COTS transceivers available today include only some of these features.

A wireless transceiver can be based on acoustic, optical or RF waves. Acoustic communication is typically used for under water communications or measuring distances based on time-of-flight measurements [2]. The disadvantages are long and variable propagation delay, high path loss, noise, and very low data rate. Also, a large external antenna is needed. In mobile networks, Doppler spread is significant reducing the data rate [8].

Optical communication [15] has low energy consumption especially in reception mode, and it can utilize very small antenna. A transmitter can be implemented by a Light Emitting Diode (LED) or a laser, and a receiver by a photo diode. However, radiation is directional and a Line-of-Sight (LOS) is required. Hence, the alignment of a transmitter to a receiver is difficult or even impossible in large-scale WSN applications.

RF communication combines the benefits of high data rate, long range and nearly omnidirectional radiation, making it the most suitable communication technology for WSNs. Disadvantages are large antenna size and higher energy consumption compared to the optical technology.

In general, an RF transceiver (radio) has four operation modes: transmit, receive, idle, and sleep. Radio is active in transmit and receive modes, when power consumption is also the highest. In idle mode, most of circuitry is shut down, but the transition to the active mode is fast. The lowest power consumption is achieved in sleep mode when all circuitry is switched off.

Most short-range radios utilized with WSNs operate in the 433 MHz, 868 MHz, 915 MHz, and 2.4 GHz license-free Industrial Scientific Medical (ISM) frequency bands. The 2.4 GHz band is the widest providing more channels, while obstacles have least effect on lower frequency bands. Depending on the frequency band and antenna type, operating range with 1 mW transmission power is from few meters to hundreds meters [7].

The characteristics of the potential commercial low power radios are summarized in Table 3.1 [7]. Microchip, Nordic Semiconductor, and Texas Instruments utilize on-chip buffers for the adaptation of a high-speed radio with a low-speed MCU. Current consumptions are specified at the lowest band and 0 dBm transmission power. The table indicates that data rate and frequency band has only a low effect on current consumption. The last two columns present the energy consumption with 3.0 V supply voltage, indicating that the radios operating at the 2.4 GHz frequency band are the most energy-efficient, which is mostly caused by their high data rates.

3.2 Computing subsystem

The central component of a platform is processor unit that forms the computing subsystem. The processor unit is typically implemented by a MCU, which integrates a processor core with program and data memories, timers, configurable I/O ports,

Table 3.1 Radio features, current consumptions, and energy efficiencies.

Radio	Data rate (kbps)	Band (MHz)	Buffer (B)	Sleep (μA)	RX (mA)	TX (mA)	RX (nJ/b)	TX (nJ/b)
MC MRF24J40	250	2400	128	2	18	22	264	216
NS nRF2401A	1000	2400	32	0.9	19.0	13.0	39	57
NS nRF24L01	2000	2400	32	0.9	12.3	11.3	17	18
NS nRF905	50	433-915	32	2.5	14.0	12.5	750	840
RFM TR1001	115.2	868	no	0.7	3.8	12	313	99
RFM TR3100	576	433	no	0.7	7.0	10	52	36
SE XE1201A	64	433	no	0.2	6.0	11.0	516	281
SE XE1203F	152.3	433-915	no	0.2	14.0	33.0	650	276
TI CC2420	250	2400	128	1	18.8	17.4	209	226
TI CC2500	500	2400	64	0.4	17.0	21.2	127	102
TI CC1000	76.8	433-915	no	0.2	9.3	10.4	406	363
TI CC1100	500	433-915	64	0.4	16.5	15.5	93	99

Manufacturers: Microchip (MC), Nordic Semiconductor (NS), sRF Monolithics (RFM), Semtech (SE), Texas Instruments (TI)

Analog-to-Digital Converter (ADC) and other peripherals. Flash memory is typically used as a program memory, while data memory consists of SRAM and Electrically Erasable Programmable Read-Only Memory (EEPROM). WSN nodes utilize typically 1-10 Million Instructions Per Second (MIPS) processing speed. Memory resources typically consists of 1-10 kB of data memory and 16-128 kB of program memory.

The characteristics of potential MCUs from different manufacturers are compared in Table 3.2. The energy-efficiencies of MCUs can be compared according to their current consumption at one MIPS processing speed. The comparison indicates that Semtech XE8802 and Texas Instruments MSP430F1611 MCUs are the most energy-efficient [7].

3.3 Sensing subsystem

There exists a large variety of low power sensors suitable for WSNs [1]. For example, sensors are available for acceleration, air pressure, humidity, illumination, infra-red, magnetic field, geographic position, and temperature. Important requirements for sensors are low power consumption and short sensing time, which determine the energy consumption of a single sensing. In addition, adequate accuracy is required within the entire temperature range. The features of some example sensors are presented in Table 3.3. Most of the sensors fulfill the requirements well.

A WSN node can also operate as a decision unit, which takes sensor readings from the WSN as input and generates action commands as output. These action commands are then transformed into actions by actuators. Besides an electric switch and a servo drive, an actuator can be a mobile robot. In order to improve the reliability of actions, the robot can be a WSN node and act based on its own sensor readings and the data of the other WSN nodes in the network [1].

Table 3.2 The comparison of the features of low power MCUs.

MCU	FLASH (kB)	SRAM (kB)	EEPROM (B)	Sleep (μA)	1 MIPS (mA)
Atmel AT89C51RE2 (8051)	128	8	0	75	7.4
Atmel ATmega103L (AVR)	128	4	4096	1	1.38
Atmel AT91FR40162S (ARM)	2048	256	0	400	0.96
Cypress CY8C29666	32	2	0	5	10
Freescale M68HC08	61	2	0	22	3.75
Microchip PIC18LF8722	128	3.9	1024	2.32	1.0
Microchip PIC24FJ128	128	8	0	21	1.6
Semtech XE8802 (CoolRisc)	22	1	0	1.9	0.3
TI MSP430F1611	48	10	0	1.3	0.33

3.4 Power subsystem

The power subsystem stores supply energy and converts it to an appropriate supply voltage level. The subsystem consists of an energy storage, a voltage regulator, and optionally an energy scavenging unit.

3.4.1 Energy Storage

The energy storage can be a non-rechargeable (primary) battery, a rechargeable (secondary) battery, or a supercapacitor [12]. Primary batteries are cheap and have the highest energy density. They are the most common power source for WSNs. Secondary batteries have lower energy density and are more expensive, but they can be recharged only 500 - 1000 times. Compared to secondary batteries, supercapacitors have lower energy density and they are more expensive. However, their lifetime is in the order of a million charging/discharging cycles. Supercapacitors are suitable to be used with an energy generator, since energy is typically generated in peaks during short periods of time, and the amount of stored energy can be relatively low.

3.4.2 Energy Scavenging

As energy storages have finite capacity, there is a requirement for self-powered devices. In energy scavenging, a node collects energy from its surrounding environment [3]. This enables more active and longer term operation by reducing the dependency on batteries or eliminates the need for them completely. Installation costs are reduced as self-powered wireless sensors do not require wires or conduits, and are therefore very easy to install. In addition, energy harvesting allows for devices to function unattended and eliminates service visits to replace batteries that can cause

Table 3.3 Features of typical sensors.

Physical quantity	Example sensor	Accuracy	Active current	Sensing time	Energy consumption
Acceleration	VTI SCA3000	1%	120 μA	10 ms	3.6 μJ
Air pressure	VTI SCP1000	150 Pa	25 μA	110 ms	8.3 μJ
Humidity	Sensorion SHT15	2%	300 μA	210 ms	190 μJ
Illumination	Avago APDS-9002	50%	2.0 mA	1.0 ms	6.0 μJ
Infrared	Fuji MS-320	-	35 μA	cont.	-
Magnetic field	Hitachi HM55B	5%	9.0 mA	30 ms	810 μJ
Position	Fastrax iTRAX03	1.0 m	32 mA	4.0 s	380 mJ
Temperature	Dallas DS620U	0.5°C	800 μA	200 ms	480 μJ

significant labor costs, especially on large networks. Finally, energy scavenging reduces environmental impact as it eliminates the need for batteries.

An energy scavenging node can use an energy storage (e.g. supercapacitors) to even differences between energy production and peaks in energy consumption. Also, batteries may still be used as the main energy source, while the energy scavenging is used to prolong the lifetime of batteries.

The two main methods for energy scavenging are radiation energy, thermal energy, and mechanical energy scavenging. In radiant energy scavenging, a node converts energy from signals send in RF spectrum. Solar cells are the most mature radiant energy scavenging technology. They can provide up to $15\,\mathrm{mW/cm^2}$ power in direct sun light at outdoor conditions [14]. However, the produced power decreases significantly in indoor conditions. In thermal energy scavenging, a node converts the heat difference between its two surfaces to energy. Even a relatively small, less than $10\,^\circ\mathrm{C}$, temperature differences can be enough for a low power WSN device. A promising method for mechanical energy scavenging is converting vibration to power with a piezoelectric conversion. Commonly occurring vibrations can provide up to $200\,\mu\mathrm{W/cm^3}$ power. Another possible energy source is air flow. For example, a self-powered device placed to a ventilation systems can to monitor air condition. The typical amount of power produced by different energy sources is summarized in Table 3.4.

As wireless technology has become more and more commonplace, energy conversion from RF waves into Direct Current (DC) power is a potential scavenging target [11]. As wireless transmitters broadcast energy to all directions, the idea is to collect some of the energy that would otherwise be wasted. There are multiple approaches to converting an RF signal to DC power depending on the desired operating parameter, such as power, efficiency, or voltage. The amount of power available for the end device depends on several factors including the source power, distance from the source, antenna gain, and conversion efficiency. The sources for RF energy harvesting can be grouped into three general categories: intentional sources, anticipated ambient sources, and unknown ambient sources.

In intentional sources, the nodes collect energy emitted from dedicated power transmitters. Thus, the availability and amount of power is controlled and engineered for the application. The intentional sources are usually comparable to the power of widely deployed RFID readers. In anticipated ambient sources, power is not explicitly generated for the sensors but can still be relied on to act as a power

Table 3.4 A comparison of typical power sources for energy scavenging.

Energy source	Power density	Duration
Solar cell (direct sun light)	$15\,\mathrm{mW/cm^2}$	Continuous
Solar cell (well illuminated room)	$10\,\mu\mathrm{W/cm^2}$	Continuous
Piezoelectric	$200\,\mu\mathrm{W/cm^3}$	Operation (e.g. button push)
Temperature difference	$40\,\mu\mathrm{W/cm^3}\,/\,5\,^\circ\mathrm{C}$	Continuous
Air flow	$380\,\mu\mathrm{W/cm^3}\,/\,5\,\mathrm{m/s}$	Continuous

source on a regular basis. An example of this are mobile phones carried by people. Thus, it can be expected that there is energy available for harvesting at popular places, e.g. bus stops. Additional examples of these sources include known radio, television, and mobile base station transmitters. In unknown ambient sources, there is no control and no prior knowledge of the exact amount of RF radiation but the sources can still provide a continual or intermittent source of power. Microwave radio links and mobile radios such as those used by police forces are examples of unknown sources.

3.4.3 Voltage Regulators

Voltage regulators can be divided into linear and switched-mode regulators. Linear regulators control the output voltage by adjusting the voltage drop across a series power transistor, which is located between the unregulated input and the regulated output voltage. This transistor conducts continuously and the energy of voltage drop is converted to heat. Thus, linear regulators are simple, cheap, small and have very low electromagnetic interferences and quiescent current [10].

Switched-mode step-down type regulators convert and input voltage to a lower output voltage by a switching transistor that is opened and closed periodically. Then, the switching current is fed to a simple coil-capacitor (L-C) filter and a diode, which average the output voltage. Thus, the energy loss of the regulator is theoretically zero. In practice, some energy is consumed in the voltage drops, resistances and leakage currents of the switching transistor, diode, coil and capacitor, and for the supply power of a controller circuitry. Compared to linear regulators, switching mode regulators are more expensive and larger, and have significantly higher electromagnetic emissions and quiescent current [10].

Switched-mode regulators provide higher energy-efficiency, when the supply current and dropout voltage are high. However, as WSNs typically use low average supply current levels, linear regulators can be more energy-efficient.For comparing the performance of linear (L) and switched-mode (S) regulators, Table 3.5 presents the features of some of the lowest power regulators available on the market.

3.5 Existing Platforms

WSN platforms have improved significantly during the last decade along with the advances in low power processing and communication technology. Still, due to the strict energy constrains, and the visions of complex networking and data fusion, it is not possible to fulfill all the requirements with the current level of technology. Thus, the platform research can be divided into two branches: high performance platforms, and low power platforms [6].

Table 3.5 Features of low-power voltage regulators.

Regulator	Type	Quiescent current (μA)	Max. load (mA)	Max. input (V)	Dropout (mV)
Maxim MAX1725	L	2.0	20	12.0	300
Maxim MAX8880	L	3.5	200	12.0	350
Microchip MCP1702	L	2.0	250	13.2	330
Minilogic ML62	L	2.8	150	10	800
SII S-1206	L	1.0	250	6.5	350
TI TPS71501	L	3.2	50	24	415
Analogic Tech AAT1112	S	42	1500	5.5	200
Microchip MCP1603	S	45	500	5.5	250
TI TPS62000	S	50	600	5.5	230

The high performance platforms have been developed for researching complex data processing and fusion in sensor nodes. The design target has been the reduction of transmitted data by efficient data processing. These platforms utilize high performance processors having at least tens of MIPS processing performance and hundreds of kilobytes program and data memories. For long-lived battery operation, their energy consumption is not adequate. However, these high power platforms can be used as a part of a WSN for data processing and data routing. Examples of the high performance platforms are Piconode [13], μAMPS [9], and Stargate [4].

Low power platforms are aiming to maximize the lifetime and minimizing the physical size of nodes. These are obtained by minimizing hardware complexity and energy consumption. These platforms are capable for performing low data rate communication and data processing required for networking and simple applications. The most essential sensor node platforms are listed in Table 3.6 [7].

Besides the COTS platforms presented in the table, a lot of research work has been conducted for developing System-on-Chip (SoC) or ASIC platforms targeting to even smaller size and higher energy-efficiency. For example, a WiseNET SoC sensor node [5] developed in Swiss Center for Electronics and Microtechnology integrates a low-power radio with CoolRISC MCU core, low-leakage memories, two ADC and power management blocks. The reception mode current consumption is only 2 mA, which is nearly one order of magnitude less than in typical low power radios. Yet, the data rate is only 25 kbps. The transmission mode current consumption at 10 dBm output power is 24 mA. The sleep mode current consumption of the radio block is 3.5 μA.

At best, low power platforms can perform various sensing tasks and they enable the extending of network lifetime to even years. However, this necessitates an energy-efficient communication protocols, which minimize the time node spends in its active modes.

Table 3.6 Comparison of existing low-power sensor node platforms.

Platform	MCU	Sensors	Radio	RF band (MHz)	Sleep (μA)	Size (mm^2)
Mica	ATmega103L	no	TR1000	915	200-300	1856
Mica2	ATmega128L	no	CC1000	433,915	17	1856
Mica2dot	ATmega128L	no	CC1000	433,915	17	492
MicaZ	ATmega128L	no	CC2420	2400	30	1856
BTnode ver3	Atmega128L	no	ZV4002 / CC1000	2400 / 433,915	3000	1890
Medusa MK-2	ATmega128L + ARM7	T,L,A	TR1000	915	27	4500
EYES node	MSP430	no	TR1001	868	5.1	2600
ScatterWeb ESB	MSP430	L,AC,A,IR	TR1001	868	8	3000
TinyNode	MSP430	no	XE1205	433-915	5.1	1200
Tmote sky	MSP430	H,T,L	CC2420	2400	5.1	2621
ProSpeckz	CY8C29666	no	CC2420	2400	330	704
TUTWSN LR	PIC18LF8722	T,L,H,A,IR	nRF905	433	31	8100
TUTWSN LE	PIC18LF8722	T,L,H,A,IR	nRF24L01	2400	24	2800

Sensors: Acceleration (A), Acoustic (AC), Humidity (H), Light (L), Passive Infra-Red (IR), Temperature (T)

References

1. Akyildiz, I.F., Kasimoglu, I.H.: Wireless sensor and actor networks: Research challenges. Elsevier Ad Hoc Networks **2**(4), 351–367 (2004)
2. Baunach, M., Kolla, R., Mühlberger, C.: Beyond theory: Development of a real world localization application as low power wsn. In: Proc. 32nd IEEE Conference on Local Computer Networks (LCN'07), pp. 872–884. Dublin, Ireland (2007)
3. Chalasani, S., Conrad, J.M.: A survey of energy harvesting sources for embedded systems. In: IEEE Southeastcon, pp. 442–447 (2008)
4. Crossbow Technology, Inc.: Stargate X-Scale processor platform. Available: http://www.xbow.com/Products/Product_pdf_files/Wireless_pdf/ 6020-0049-01_B_STARGATE.pdf (2004)
5. Enz, C.C., El-Hoiydi, A., Decotignie, J.D., Peiris, V.: WiseNET: An ultralow-power wireless sensor network solution. Computer **37**(8), 62–70 (2004)
6. Hill, J., Horton, M., Kling, R., Krishnamurthy, L.: Wireless sensor networks: The platforms enabling wireless sensor networks. Communications of the ACM **6**(47), 41–46 (2004)
7. Kohvakka, M.: Medium access control and hardware prototype designs for low-energy wireless sensor networks. Ph.D. thesis, Tampere University of Technology, Tampere, Finland (2009)
8. Kong, J., Cui, J.H., Wu, D., Gerla, M.: Building underwater ad-hoc networks and sensor networks for large scale real-time aquatic applications. In: Proc. IEEE Military Communications Conference (MILCOM'05), vol. 3, pp. 1535–1541. Atlantic City, NJ, USA (2005)
9. Min, R., Bhardwaj, M., Cho, S.H., Ickes, N., Shih, E., Sinha, A., Wang, A., Chandrakasan, A.: Energy-centric enabling technologies for wireless sensor networks. IEEE Wireless Communications **9**(4), 28–39 (2002)
10. Mozar, S.: An evaluation of low cost power supply alternatives for high volume consumer products. In: Proc. 10th Int. Symposium on Consumer Electronics (ISCE'06), pp. 1–4. St. Petersburg, Russia (2006)
11. Paradiso, J., Starner, T.: Energy scavenging for mobile and wireless electronics. Pervasive Computing, IEEE **4**(1), 18–27 (2005). DOI 10.1109/MPRV.2005.9

12. Pitcher, G.: If the cap fits... New Electronics pp. 25–26 (2006)
13. Reason, J.M., Rabaey, J.M.: A study of energy consumption and reliability in a multi-hop sensor network. ACM SIGMOBILE Mobile Computing and Communications Review **8**(1), 84–97 (2004)
14. Roundy, S., Wright, P.K., Rabaey, J.: A study of low level vibrations as a power source for wireless sensor nodes. Computer Communications **26**(11), 1131–1144 (2003)
15. Wolf, M., Kress, D.: Short-range wireless infrared transmission: the link budget compared to RF. IEEE Wireless Communications Magazine **10**(2), 8–14 (2003)

Chapter 4
Communication Protocols

Network protocols are typically divided into several distinct layers according to their responsibilities, which together form a protocol stack. Each layer has precisely defined interfaces, which permits flexible updates and changes in the software and hardware implementations in a modular manner. A WSN communication protocol stack comprises typically MAC, routing, and transport protocols. This chapter describes the design concepts and typical approaches in these protocol layers.

4.1 Medium Access Control Features and Services

The MAC sublayer is the lowest part of data link layer and it operates on top of the physical layer. A MAC protocol manages radio transmissions and receptions on a shared wireless medium and provides connections for overlying routing protocol. Hence, it has a very high effect on network performance and energy consumption.

4.1.1 Low-Energy MAC Design

As WSNs are designed to operate in large geographic areas, the energy consumption of data transmissions is reduced by forwarding data in the network by several low energy hops (multi-hop routing). According to a distance-dependent path loss model [18], the mean path loss ($\overline{PL}$) increases exponentially with distance (d) as

$$\overline{PL}(d) \propto \left(\frac{d}{d_0}\right)^n, \tag{4.1}$$

where n is the path loss exponent, d_0 is the reference distance, and d is the actual distance between the transmitter and the receiver. The value of the path loss exponent is typically between 2 and 3.5 depending on the operation environment [18, 16]. Thus, the reduction of a hop length to a half reduces the required trans-

27

mission power to between 1/4 and 1/12. Yet, this model considers only the radiated transmission power and ignores the static power consumption of the transmitter and receiver circuitry, which dominate the power consumption at short hop distances. Thus, the advantage of multi-hop routing is limited [26, 3].

For reliability and energy-efficiency, a MAC protocol should minimize:

- Idle listening: Idle listening occurs when a node is actively receiving a channel, but there is no meaningful activity on the channel resulting wasted energy.
- Collisions: When to nodes transmit simultaneously at the same frequency channel their transmissions collide in the overlapping area of their transmission ranges. In this area the received data is most probably corrupted causing useless receive cost at the destination node and useless transmit cost at the source node.
- Overhearing: An unicast transmission on a shared wireless broadcast medium may cause other nodes than the intended destination to receive a data packet, which is most probably useless to them and consumes unnecessarily energy.
- Protocol overhead: As WSNs utilize relatively short packets, protocol headers and trailers may be even longer than the actual transmitted payload data causing significant energy consumption. In addition, the energy consumption caused by control packet exchange, such as Request To Send (RTS)/Clear-To-Send (CTS) may be very significant.

Furthermore, the operating mode change transients may dissipate a lot of energy, e.g. from sleep mode to transmit mode, which should be considered, as well [22].

As a principle, the highest energy-efficiency is achieved, when a source and a destination node are activated and tuned on a correct RF channel simultaneously, while other nodes remain in sleep mode. This is very difficult in large and resource constrained WSNs, where the network topology is constantly changing. The design requirements can be reached only by careful cross layer design, where each protocol layer is designed with an accurate knowledge of the influences to lower protocol layers and finally to node energy and memory consumption and network performance.

4.1.2 MAC Technologies

MAC protocols can be categorized into contention and contention-free protocols. In contention protocols, nodes compete for a shared channel, while trying to avoid frame collisions.

As the power consumption of the low power radios in the reception mode is high, the energy-efficiency of the conventional MAC approaches is not adequate for the low energy WSN as such. Further energy saving is achieved by *duty cycling*: time is divided into a short active period and a long sleep period, which are repeated consecutively. These low duty-cycle protocols can also be divided into two categories:

unsynchronized and synchronized protocols, according to the synchronization of data exchanges.

ALOHA [17] is the simplest contention protocol, where nodes transmit data without coordination. Slotted ALOHA reduces collisions by dividing time into slots and transmitting data on the slot boundaries only. Carrier Sense Multiple Access (CSMA) further reduces collisions and improves achievable throughput by checking channel activity prior to transmissions and avoiding transmission during busy channel situations.

Carrier Sense Multiple Access with Collision Avoidance (CSMA/CA) [2] is a modification of CSMA, which reduces the congestion on a channel by deferring a transmission for a random interval (contention window). The contention window is increased if the channel is sensed to be busy (backoff), thus allowing the MAC to adjust to the network conditions. Still, collisions may occur due to a hidden node problem: nodes separated by two hops may not detect each other, and their transmissions may collide on a receiver that is located between the nodes. The hidden node collisions can be significantly reduced by performing a RTS/CTS handshaking prior to a data transmission. Therefore, the handshaking is defined as an option in many CSMA/CA-based protocols. While contention-based protocols work well under low traffic loads, their performance and reliability degrades drastically under higher loads because of collisions and retransmissions.

In contention-free protocols, nodes get unique time slots or frequency channels for transmissions. Ideally, collisions are eliminated. TDMA divides time into numerous slots, where only one node is allowed to transmit on each slot. Other alternatives are Frequency Division Multiple Access (FDMA) and Code Division Multiple Access (CDMA) , which provide contention-free operation by separate frequency channels and spreading codes, respectively. Contention-free protocols achieve high performance and reliability regardless of the traffic load. Yet, the bandwidth must be reserved in advance, which increases control traffic overhead.

4.1.3 Unsynchronized Low Duty-Cycle MAC Protocols

Unsynchronized low duty-cycle MAC protocols [15] are based on a Low Power Listening (LPL) mechanism, where nodes poll channel asynchronously to test for possible traffic, as presented in Fig. 4.1. Transmissions are preceded with a preamble that is longer than the channel-polling interval. Hence, the preamble part acts like a wake up signal. If a busy channel is detected, nodes begin to listen to the channel until a data packet is received or a time-out occurs. Bluetooth Low Energy , DASH7 , and protocols are based on the LPL mechanism.

The drawback of the basic LPL mechanism is that the transmission and reception of long preamble increases energy consumption significantly. Therefore, several variations are proposed to reduce the preamble energy. For example, in X-MAC [1], a sender transmits multiple short preambles with the address of the intended receiver. Each preamble is followed by a short reception period. Upon receiving

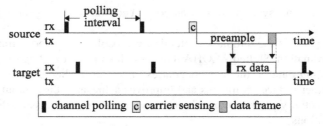

Fig. 4.1 Operation of unsynchronized low duty-cycle protocols.

a preamble, the destination node sends an acknowledgment (ACK) between the preambles. Other nodes can enter early a sleep mode for reducing overhearing. After receiving the ACK, the source node begins the transmission of a data frame.

The preamble can be eliminated completely by utilizing an additional transceiver referred to as a wake-up radio [4]. The wake-up radio mechanism is based on the assumption that the listen mode of the wake-up radio is ultra low power and it can be active constantly. At the same time, the normal data radio is in the sleep mode as long as packet transmission or reception is not required. The wake-up radio protocols are successful in avoiding overhearing and idle listening in the data radio. Their major problems are the energy consumption and cost of the wake-up radio. In addition, the difference in the transmission ranges between data and wake-up radio may pose significant problems.

Unsynchronized protocols are relatively simple and robust, and require a small amount of memory compared to synchronized protocols. A general drawback is rather high overhearing, since each node must receive at least the beginning of each frame transmitted within radio range. Thus, they suit best for relatively simple WSNs utilizing very low data rates. Unsynchronized protocols tolerate dynamics in networks, but their energy-efficiency is limited by the channel sampling mechanism [25].

4.1.4 Synchronized Low Duty-Cycle MAC Protocols

Synchronized low duty-cycle MAC protocols utilize scheduling to ensure that nodes agree on the data exchange times. Due to the synchronized operation, nodes know the exact moments of active periods in advance, thus eliminating the need of long preambles. As a global synchronization is very difficult in large networks, the synchronization is often realized by receiving a beacon frame from one or more neighbor nodes, as shown in Fig. 4.2. The beacon frame includes synchronization and status information, such as the duration of the active period and the time between active periods. After a beacon, nodes exchange frames during the active period. The active period is followed by a sleep period to save energy. Together, the active period and the sleep period are referred to as an *access cycle*. The access cycle is repeated periodically.

For establishing the synchronized operation, neighboring nodes are typically discovered by a network scan. The network scan means a long-term reception of frequency channels for receiving beacons from neighbors, since their schedules and frequency channels are unknown. Clearly, this is energy-hungry. However, the synchronized operation after the network scan is very energy-efficient [25].

While the channel access in the unsynchronized protocols is usually contention-based, the synchronized MAC protocols use either contention-based, contention-free, or hybrid channel access mechanism.

Sensor-MAC (S-MAC) [24] utilizes purely contention-based channel access by using CSMA/CA with RTS/CTS mechanism. The protocol utilizes a fixed active period length and an adjustable, network specific wake up period. Neighboring nodes may coordinate their active periods to occur simultaneously to form virtual clusters. At the beginning of an active period nodes wake up and exchange synchronization (SYNC) frames for synchronizing their operation. The fixed access cycle length causes idle listening, which decreases energy efficiency. T-MAC [20] is a variation of the S-MAC that improves the energy-efficiency by adjusting the active period according to traffic. It utilizes a short listening window after the CTS phase and each frame exchange. If no activity occurs during the listening window, node returns to sleep mode.

IEEE 802.15.4 [5] standard defines a MAC layer that can use both contention-based and contention-free channel access. It operates on beacon-enabled and non-beacon modes. In the non-beacon mode, a protocol is based on a simple CSMA/CA. Energy-efficient synchronized low duty-cycle operation is provided by the beacon-enabled mode, where all communications are performed in a superframe structure. The superframe is divided into three parts: the beacon, Contention Access Period (CAP) and Contention-Free Period (CFP). CAP is a mandatory part of a superframe during which channel is accessed using a slotted CSMA/CA scheme. CFP is an optional feature of IEEE 802.15.4 MAC, in which a channel access is performed in dedicated time slots. CFP can be utilized only for a direct communication with a PAN coordinator. Thus, its applicability and benefits are very limited in multi-hop networks. The cluster-tree type IEEE 802.15.4 network can provide comparably good energy-efficiency in static and sparse networks. The hidden node

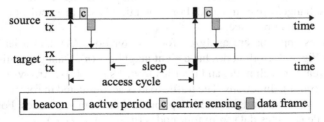

Fig. 4.2 Operation of synchronized low duty-cycle protocols.

problem reduces performance in dense networks, since any handshaking prior to transmissions is not used.

TUTWSN MAC [11, 9] is another example of protocol that uses both contention-based and contention-free channel access. The superframe structure is similar to the IEEE 802.15.4. However, instead of using carrier sensing, CAP and CFP are divided into fixed time-slots. To allow implementation on the simplest radios without carrier sensing capabilities, TUTWSN MAC uses slotted ALOHA on CAP. Each time slot is further divided into two subslots, first subslot is for data frame and the following subslot is for acknowledgment. The use of contention free slots is preferred as it eliminates collisions and thus increases reliability. The CAP is used only for joining a cluster and requesting reservations on CFP.

In the synchronized low duty-cycle protocols, the major advantage is that a sender knows a receiver's wake up time in advance and thus can transmit efficiently. In dynamic networks, synchronized links are short-lived and new neighbors need to be searched frequently, which increases energy consumption rapidly. In contention based protocols, a major disadvantage is the energy cost of receiving an entire active period[15]. Contention-free protocols have better energy-efficiency in stationary networks, but their performance reduces rapidly as network dynamics increases.

4.1.5 Performance Comparison

This section analyzes the performance of the low-energy MAC protocols. The results are based on the models presented in [9] and [25]. The models have the following assumptions:

- Each sensor node measures one sensor sample and forwards it to a next-hop node during one data generation interval,
- Each data frame is followed by an acknowledgment for fair comparison,
- There are no transmission errors nor collisions,
- There is no contention, and carrier sense attempts produce an idle result,
- The power consumption of idle listening equals to the reception mode power, and
- The active time of MCU equals to the active time of radio.

Therefore, the performance models focus on the power consumption of the channel access mechanisms, while the effects of data processing, contention, and control frame exchanges are eliminated. For contention based protocols, the results are slightly better than in practice.

Energy consumptions are analyzed for a router node (A), and a leaf node (B) presented in Fig. 4.3. Both nodes have eight neighbors (n). Data generation interval (T_{DATA}) is equal for each node, and it varies from 1 s to 1000 s. Arrows in the figure indicate data routing directions. The traffic load is accumulated in routers, since they transmit their own data and the multi-hop routed data from n_{DL} nodes. For example, the router node C routes data from four nodes ($n_{DL} = A, B, D, E$).

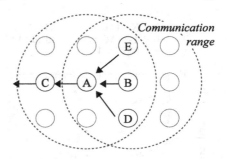

Fig. 4.3 Network topology
in WSN MAC performance
evaluation.

Average power consumption (P) of a protocol is calculated by normalized transmission (t_{TX}) and reception (t_{RX}) activities and their power consumptions as

$$P = t_{TX}P_{TX} + t_{RX}P_{RX} + (1 - t_{TX} - t_{RX})P_S. \qquad (4.2)$$

The normalized activity is determined by dividing the duration of an activity by the interval of the activity resulting in a percentage value of the activity. Data exchanges are normalized by T_{DATA} during which all nodes in the network generate exactly one data frame. Similarly, the transmission and reception activity for maintaining synchronization is normalized by T_{SYNC}.

The performance analysis assumes the commonly used TI CC2420 transceiver and Atmel ATmega128L MCU. As the transceiver and MCU constitute the majority of power consumption, other power consumption sources are ignored in the analysis.

Other analysis parameters are as follows. For fair comparison, all protocols use 8 B control frame (Beacon/ACK/RTS/CTS) length and 32 B data frame length. In IEEE 802.15.4 and T-MAC protocols, 2 ms average contention window is used, which conforms the default settings of IEEE 802.15.4 when there is no collisions. T-MAC uses 90 s synchronization interval, while TUTWSN utilizes 2 ALOHA slots per CAP. For realistic results, 20 ppm maximum clock drift was assumed. The clock drift reflects the inaccuracy of the timing crystals, which must be compensated by extra listening of the channel as a neighbor node might begin its transmission earlier or later than expected.

The power consumption with the analyzed protocols is presented in Fig. 4.4. In this analysis, the access cycle length (synchronized protocols) and the listening interval (LPL-based X-MAC protocol) are adjusted for lowest possible energy consumption. In the synchronized protocols, the optimal access cycle length ranges from 2 s to 2000 s as the data generation interval ranges from 1 s to 1000 s. The longer access cycle saves energy, as beacons need to sent less frequently, while shortening the access cycle gives more capacity. The active period length is long enough to allow one data transmission. In the unsynchronized X-MAC, the channel listening interval ranges from 0.1 s to 3.3 s. Long listening intervals make LPL-based protocols energy inefficient, because the transmission time of the preamble signal is relative to the channel listening interval.

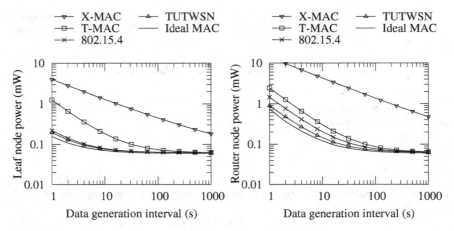

Fig. 4.4 Optimal power consumption of MAC protocols.

Clearly, the synchronized protocols are potentially the most energy efficient choices for a WSN MAC. However, the energy efficiency has a trade-off with latency. The maximum per hop forwarding delay, assuming no packet errors and collisions, is the same as the access cycle length or channel listening interval. The average per hop delay is half of the maximum delay.

Fig 4.5 shows the power consumption of the protocols with comparable delays. On each case, the access cycle length is limited to 1 s. X-MAC uses 0.1 s-0.8 s between 1 s-64 s data generation intervals, as it allows better energy-efficiency.

TUTWSN and IEEE 802.15.4 are the most energy-efficient choices for leaf nodes, as they minimize idle listening. In T-MAC, the power consumption of a leaf

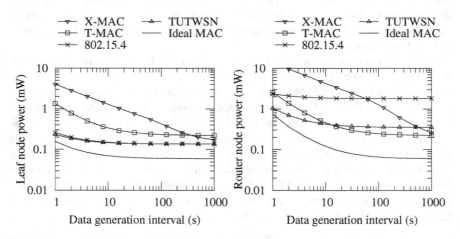

Fig. 4.5 Power consumption of MAC protocols, when maximum per-hop delay is 1 s.

is high, because the protocol does not make a distinction between router and leaf nodes. In this comparison, 802.15.4 has the highest router node power consumption among synchronized MACs as its active period (CAP) length is fixed, which causes a lot of idle listening. T-MAC and TUTWSN are more energy-efficient because they have mechanisms to minimize idle listening. T-MAC adjusts the active period length based on the traffic, whereas TUTWSN prefers contention free channel access and adjusts the amount of reserved slots dynamically according to the traffic.

The LPL-based X-MAC protocol is the least energy-efficient when network traffic is high, but has the lowest power consumption when data generation interval is low as a node uses energy only to transmit data and does not have to maintain synchronization. T-MAC is the most energy-efficient synchronized protocol when data generation interval is long, because its energy-efficient synchronization mechanism. In IEEE 802.15.4 and TUTWSN, beacon frames are transmitted every access cycle, which means that at long data generation interval, the power consumed due to beacon reception dominates.

The results indicate clearly that there is no single purpose, fit-for-all low-energy WSN MAC. The optimal MAC depends on the required delays and data generation intervals. For example, a synchronized MAC can be selected over a LPL MAC even in very low traffic networks, if the delay is not critical. This is the case e.g. in environmental networks in which samples need to be collected only once per hour. Another consideration is the role of the nodes. If the backbone network consists of router nodes that are mains powered, IEEE 802.15.4 would be a good choice as its leaf nodes have very low energy consumption.

4.2 Routing Paradigms and Techniques

As WSNs are designed to operate in large geographic areas, forwarding data directly to the target node would not be feasible as the required transmission energy increases proportionally to the square of the distance. Therefore, data is routed along several hops. A routing layer operates on top of the MAC layer. As several alternative routes to a destination node may exist, the routing decision has a significant effect on load balancing, end-to-end reliability, and latency. Furthermore, the route construction and maintenance methods used in a routing protocol determine energy-efficiency and mobility support. Due to resource constraints, WSN routing protocols often also combine transport layer functionality.

The basic service for a routing protocol is the multihop forwarding a packet from a source to a destination. However, a routing protocol may also provide:

- *QoS support* allowing route selection based on different QoS-metrics, such as end-to-end latency, reliability, or energy usage,
- *Multicast and broadcast support* allowing efficient packet delivery to several nodes at once, and
- *Mobility support* enabling source, intermediate, and/or destination node mobility

Overall, the supported services are largely limited by the used routing paradigm and technology. WSN routing protocols can be classified based on their operation as node-centric, data-centric, location-based, multipath, or cost-based [6, 13] as shown in Fig. 4.6. The classes are not exclusive as a routing protocol may be both data-centric and query based, while having features seen in location based protocols.

4.2.1 Node-centric Routing

Node-centric approach is the traditional approach used in the computer networks in which nodes are addressed with globally unique identifiers. While the paradigm allows compatibility with the existing protocols, the requirement of required unique addressing is challenging in WSNs. Due to the large network size and error prone nature of sensor nodes, a decentralized address maintenance is preferred. However, as a network may partition or network segments may join, ensuring a consistent addressing scheme involves a lot of messaging and is energy-consuming.

Node-centric protocols typically rely on routing tables containing an entry for each route identified by destination address and next hop node for the target. The routing table may be constructed *proactively* by discovering routes to all potential targets, but this increases memory requirements and would not be practical in large networks. Instead, the node-centric protocols designed for ad-hoc wireless networks, such as Dynamic Source Routing (DSR) [7], or Ad-hoc On-demand Distance Vector routing (AODV), use *on-demand* (reactive) approach in which routes are constructed only when needed. The drawback compared to the proactive approach is the route construction delay when sending first packets.

4.2.2 Data-centric Routing

As WSNs are inherently data oriented, the data centric routing is a more natural paradigm than the node-centric approach. In data-centric routing, data is routed

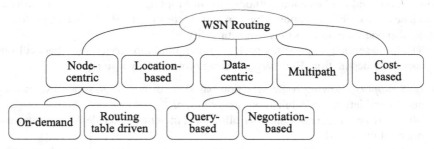

Fig. 4.6 WSN routing paradigms.

based on its content rather than using sender or receiver identifiers. As the data-centric routing is already content aware, data-aggregation can be naturally performed.

Data centric routing may take interest based, negotiation based, or query based approaches. In the interest based approach, a sink node request data from the network by sending a request describing the data it wants to every node in the network [6]. A node forwards the interest and directs its routing tree toward the sink node as shown in Fig. 4.7. Then, nodes that fulfill the requirements as defined in the interest start transmitting data to the sink. Although the route construction is proactive, the interest based routing is scalable as the number of sinks (data consumers) is low compared to the number of nodes (data sources).

Negotiation-based protocols exchange negotiation messages before actual data transmission takes place [10]. This saves energy, as a node can determine during the negotiation that the actual data is not needed. For negotiation protocols to be useful, the negotiation overhead and data descriptor sizes must be smaller than the actual data. Query based routing protocols request a specific information from the network. A query might be expressed with a high level language such as Structured Query Language (SQL). For example, a query might request "average temperature around area x,y during the last hour". The query can be routed via a random walk or directed at a certain region [12]. After the query has been resolved, the result is transmitted back to the source.

4.2.3 Location-based Routing

Location-based routing uses geographic location information to make routing decision. The approach is natural to WSNs, as sensor measurements usually relate to a specific location. A basic principle in the geographic routing is to select a next hop neighbor that is closer to the target node than a forwarding node. However, a problem with such greedy forwarding is that routing may fail due to hole in the network. Proposed solutions to the problem include switching to a different mode when a hole is detected, such as in Greedy Perimeter Stateless Routing (GPSR) [8] where packet is routed around a hole according to right-hand rule. Another method is to express packet's route with mathematical formula as proposed in Trajectory Based

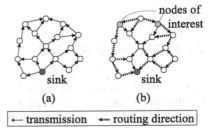

Fig. 4.7 Interest based routing. **a** sink advertises its interests to the network. **b** nodes matching the interest transmit data to the sink.

Forwarding (TBF) [14]. Nodes forward packet to a neighbor that is closest to the defined route (trajectory). With the help of global knowledge of the network, a route that avoids holes can be selected.

The most significant benefit of the location-based routing is its scalability. Routing tables or a global knowledge of the network topology is not typically required, which reduces both data memory requirements and routing overhead. Also, geographic routing usually tolerates source and intermediate node mobility. However, determining the position for each node can be problematic. The use of positioning chips such as GPS increases the price and energy consumption, while manual configuration is not suitable for large scale networks.

4.2.4 Multipath Routing

In the multipath routing, a packet traverses from a source node to a target node via several paths. The main goal is to increase reliability, as a packet can be received via an alternative path even if the routing in some path fails. However, the multipath routing has a trade-off between the reliability and energy, as it increases network load and energy usage due to the extra transmissions.

Flooding packet to every node in the network is the simplest case of multipath routing. In flooding, each node forwards a new flood packet to all of its neighbors. To suppress duplicates, already received flood packets are not forwarded. Flooding is commonly used during the setup phase of several WSN routing protocols, but is not used for routing as such because packets can easily congest network and thus decrease reliability.

Controlled multipath routing algorithms limit the number of alternative routes. For example, in gradient broadcast [23] data is forwarded along an interleaved mesh. Each packet is assigned with a budget that is initialized by the source node. The budget consists of the minimum cost to send a packet to the sink and an additional credit. When a node receives the packet, it compares remaining budget against the cost required to forward the packet to the sink. If the cost is smaller or equal than the budget, the node forwards the packet. As the credit increases the budget, it allows forwarding the packet along other than minimum cost paths. Thus, the credit determines the amount of redundancy for the packet and has a trade-off between used energy and reliability. If credit is zero, packet must be forwarded along minimum cost path.

4.2.5 Cost-based Routing

In cost-based routing, each node is assigned with a cost value that is relative to the distance between a node and a sink. The cost may be calculated from an any metric, e.g. the number of hops or the required energy to forward a packet to the sink. The benefit of the cost-based routing is that the knowledge of forwarding path states is not required: a node forwards its data by sending it to any neighbor that has lower cost. The drawback is that the routes must be created proactively. Also, although data to the sink is forwarded efficiently, another routing mechanism, such as flooding, must be used for data traveling in the other direction. However, the trade-off can be acceptable since most of the traffic is usually toward the sink.

4.3 Transport Protocols

A transport protocol regulates end-to-end traffic flow within a network. Generally, it has services for:

- *End-to-end reliability* to ensure that a packet is not lost and performs retransmissions if necessary,
- *Congestion and flow control* to avoid packet drops due to traffic congestion, and
- *Fragmentation* to enable transmission of large contents.

It should be noted that due to the cross-layer design of WSN protocols, some of these functionalities may be implemented on routing layer and a separate transport protocol is not always necessary.

Pump-Slowly, Fetch-Quickly (PSFQ) [21] combines the functionality of transport and routing layers to achieve a low communication cost. Data is transmitted with relatively slow speed by delaying forwarding with two configured time values T_{min} and T_{max}. In broadcast networks, the T_{min} parameter allows a node to receive a frame multiple times. A node then evaluates the necessity to forward the frame based on how many times it was received. If a sequence number gap in a received frame is detected, PSFQ uses a negative acknowledgment to request all missed frames. The frame is requested in less than T_{min}, which allows reducing latency on error situations.

SPEED [19] is a routing protocol that combines non-deterministic location-based forwarding with inbuilt congestion control mechanism and soft latency guarantees. The protocol does not guarantee strict limit for latency, but defines an end-to-end delay that is proportional to the distance between source and destination nodes. Thus, it maintains a certain delivery speed. In SPEED, the next hop is selected randomly among the neighbors with the probability that is proportional to the link speed. Only the nodes that advance towards the target and meet the delivery time can be selected. The link speed is calculated by dividing the distance between nodes (obtained with the geographic location information) by measured link delay. The next hop selection is combined with feedback received from neighbors. If a node cannot forward a packet due to congestion or a hole in the network, it sends a backpressure beacon, which reduces the forwarding probability to that node.

References

1. Buettner, M., Yee, G., Anderson, E., Han, R.: X-MAC: A short preamble MAC protocol for duty-cycled wireless sensor networks. In: Proc. 4th ACM Conf. Embedded Networked Sensor Systems (SenSys'06), pp. 307–320. Boulder, Colorado, USA (2006)
2. Colvin, A.: CSMA with collision avoidance. Computer Communications **6**(5), 227–235 (1983)
3. Deng, J., Han, Y.S., Chen, P.N., Varshney, P.K.: Optimal transmission range for wireless ad hoc networks based on energy efficiency. IEEE Transactions on Communications **55**(9), 1772–1782 (2007)
4. Guo, C., Zhong, L., Rabaey, J.: Low power distributed MAC for ad hoc sensor radio networks. In: Global Telecommunications Conf. (GLOBECOM'01), vol. 5, pp. 2944–2948. San Antonio, TX, USA (2001)
5. IEEE 802.15.4: IEEE Standard for Information Technology—Telecommunications and Information Exchange Between Systems—Local and Metropolitan Area Networks—Specific Requirements—Part 15.4: Wireless Medium Access Control (MAC) and Physical Layer (PHY) Specifications for Low-Rate Wireless Personal Area Networks (LR-WPAN) (2006)
6. Al Karaki, J.N., Kamal, A.E.: Routing techniques in wireless sensor networks: A survey. IEEE Wireless Communications **11**(6), 6–28 (2004)
7. Karl, H., Willig, A.: Protocols and Architectures for Wireless Sensor Networks. John Wiley & Sons Ltd (2005)
8. Karp, B., Kung, H.T.: GPSR: Greedy perimeter stateless routing for wireless networks. In: Proc. 6th annual Int'l Conf. on mobile computing and networking (MobiCom'00), pp. 243–254. Boston, MA, USA (2000)
9. Kohvakka, M.: Medium access control and hardware prototype designs for low-energy wireless sensor networks. Ph.D. thesis, Tampere University of Technology, Tampere, Finland (2009)
10. Kulik, J., Heinzelman, W., Balakrishnan, H.: Negotiation-based protocols for disseminating information in wireless sensor networks. Kluwer Wireless Networks **8**(2), 169–185 (2002)
11. Kuorilehto, M., Kohvakka, M., Suhonen, J., Hämäläinen, P., Hännikäinen, M., Hämäläinen, T.D.: Ultra-Low Energy Wireless Sensor Networks in Practice - Theory, Realization and Deployment. John Wiley & Sons Ltd (2007)
12. Liu, J., Zhao, F., Petrovic, D.: Information-directed routing in ad hoc sensor networks. IEEE Journal on Selected Areas in Communications **23**(4), 851–861 (2005)
13. Niculescu, D.: Communication paradigms for sensor networks. IEEE Communications Magazine **43**(3), 116–122 (2005)
14. Niculescu, D., Nath, B.: Trajectory based forwarding and its applications. In: Proc. 9th annual Int'l Conf. on Mobile computing and networking (MobiCom'03), pp. 260–272. San Diego, CA, USA (2003)
15. Polastre, J., Hill, J., Culler, D.: Versatile low power media access for wireless sensor networks. In: Proc. 2nd Internation Conf. on Embedded Networked Sensor Systems (Sensys'04), pp. 95–107. Baltimore, MD, USA (2004)
16. Rappaport, T.: Wireless Communications - Principles and Practice, 2 edn., chap. 1. Prentice Hall (1996)
17. Roberts, L.: ALOHA packet system with and without slots and capture. ACM SIGCOMM Computer Communication Review **5**(2), 28–42 (1975)
18. Seidel, S., Rappaport, T.: 914 MHz path loss prediction models for indoor wireless communications in multifloored buildings. IEEE Trans. Antennas and Propagation **40**(2), 207–217 (1992)
19. Tian He, Stankovic, J.A., Lu, C., Abdelzaher, T.: SPEED: A stateless protocol for real-time communication in sensor networks. In: Proc. 23rd Int'l Conf. on Distributed Computing Systems, pp. 46–55. Providence, RI, USA (2003)
20. van Dam, T., Langendoen, K.: An adaptive energy-efficient MAC protocol for wireless sensor networks. In: Proc. 1st Int'l Conf. on Embedded Networked Sensor Systems (Sensys'03), pp. 171–180. Los Angeles, CA, USA (2003)

21. Wan, C.Y., Campbell, A.T., Krishnamurthy, L.: Pump-slowly, fetch-quickly (PSFQ): A reliable transport protocol for sensor networks. IEEE Journal on Selected Areas in Communications **23**(4), 862–872 (2005)
22. Warneke, B., Last, M., Leibowitz, B., , Pister, K.S.J.: Smart dust: Communicating with a cubic-millimeter computer. Computer **34**(1), 43–51 (2001)
23. Ye, F., Zhong, G., Lu, S., Zhang, L.: GRAdient broadcast: a robust data delivery protocol for large scale sensor networks. Kluwer Wireless Networks **11**(3), 285–298 (2005)
24. Ye, W., Heidemann, J., Estrin, D.: An energy-efficient MAC protocol for wireless sensor networks. In: Proc. 21st Annual Joint Conf. of the IEEE Computer and Communications Societies (INFOCOM'02), vol. 3, pp. 1567–1576. New York, NY, USA (2002)
25. Yoon, S.: Power management in wireless sensor networks. North Carolina State University, PhD Thesis (2007)
26. Zhang, R., Gorce, J.M.: Optimal transmission range for minimum energy consumption in wireless sensor networks. In: Proc. IEEE Wireless Communications and Networking Conference (WCNC'08), pp. 757–762. Budapest, Hungary (2008)

Chapter 5
Software and Middleware Services

The software components in WSNs include sensor operating systems and middleware as shown in Fig. 5.1. The purpose of these is to ease the application development by providing network access and allowing support for heterogeneous platforms with hardware abstraction.

5.1 Sensor Operating Systems

An operating system eases the use of system resources. Its main functions are the concurrent execution and communication of multiple applications, access and management of Input/Output (I/O) devices, permanent data storage, and control and protection of system access between multiple users [28].

While most of the typical OS functions also apply to WSNs, not all of them are significant. Instead of the human centered perspective, the main objective of a WSN OS is to make application and system development easier. The key requirements for WSN Operating Systems (OSs) can be summarized as [14]:

- *Small memory footprint*: Due to the resource constraints , an operating system should be as small as possible to leave space for applications.

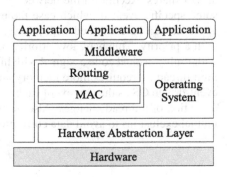

Fig. 5.1 Software architecture of a WSN node.

- *Realtime operation*: Sensor networks are inherently coupled with the real world, which sets timing constraints to the operation. In addition, network protocols are often time-sensitive. Both of these require realtime support to make right actions at the right time.
- *Energy management*: As sensor nodes may operate on batteries, an OS should operate efficiently to reduce its energy overhead by e.g. shutting down unused peripherals.
- *Memory management*: The scarce data memory should be efficiently divided among applications.
- *Hardware abstraction*: In order to provide coherent interfaces for applications, an OS needs to abstract heterogeneous node platforms. This way, a developer can write applications using a single Application Programming Interface (API).
- *Concurrency*: WSNs have high degree of concurrency. Sensing is not limited to one application but can involve multiple physical sensors. Also, a node must be able to forward and process (e.g. via aggregation) several concurrent data flows. Thus, basic services that allow concurrency are essential in an OS.

5.1.1 Implementation Approaches

Sensor operating systems have two basic implementation approaches: event-driven and preemptive multithreading [14]. Both approaches allow real-time responses but differ in how they are programmed.

In an event-driven OS, processes are often referred to as tasks. A task is activated when an event occurs and is run to completion, as shown in Figure 5.2(a). An event might indicate e.g. the reception of data from a transceiver, the expiration of a timer, or finished sensing. The main benefit of the approach is extremely low overhead: the event handler can be invoked with a direct function call without any context (hardware registers, software stack, etc.) saving. However, the approach is not suitable for long running computations as they block execution of other events. A programmer can work around this issue by manually slicing the computation to little pieces but this increases the complexity of the software.

Preemptive multithreading allows a thread (or process) to run constantly, while an OS shares execution time between active threads. A thread may decide to wait some specific event in which case the execution halts until the event occurs. The approach can be combined with priority scheduling where each thread is assigned with a priority. This way, low priority background computation does not interfere with timing critical code assigned with high priority. As a drawback, the preemptive multithreading requires more memory and has higher computational overhead than event-driven OS as the context of each thread must be stored and switched when the execution of a thread changes. Also, sharing data structures and resources requires mutual exclusions to avoid data corruption.

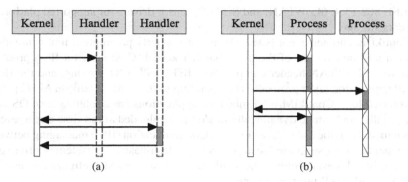

Fig. 5.2 WSN operating system multitasking principles. **a** In event-driven approach task is run to completion. **b** Preemptive multitasking allows suspending a process to run another process.

5.1.2 Existing Operating Systems

Several existing operating systems are suitable for resource constrained WSNs. These include both OSs that are generally targeted at embedded devices but run on typical WSN hardware, and OSs that are specifically designed for WSNs. The main advantage of the dedicated WSN OSs is that they often include application and network protocol functionality to ease sensor data collection and processing.

Examples of generic embedded OSs are LiteOS, FreeRTOS, and μC/OS. LiteOS is a Unix-like open source operating system that uses preemptive multitasking. The use of conventional threads instead of events is intended to simplify sensor programming with the familiarity of the traditional programming techniques. The OS features plug-in routing stack, extensive event logging support, and built-in hierarchical file system. FreeRTOS is another open source OS but has also a commercially supported version, OpenRTOS. FreeTOS supports both preemptive and co-operative multitasking where a thread voluntarily releases its execution on a blocking OS call. μC/OS is commercial operating system that is specifically targeted at safety critical applications, such as used in aviation or medical systems.

TinyOS, Contiki, and MANTIS (MultimodAl NeTwork's for In-situ Sensors) are open source OSs specifically designed for WSNs. TinyOS [11] is an event-driven OS where a task is run to in respect to other tasks but can be interrupted by events. In this sense, the OS takes care of concurrency although programmer controlled preemptive threading is possible with extensions. TinyOS was originally designed for Berkeley motes, but has been later ported to many other platforms. TinyOS is programmed with nesC language, a derivative of C, that puts emphasis on modular, component-based programming. For performance reasons, the components are statically linked to each other via their interfaces. The compile process translates nesC code to C code which is then compiled with a native compiler. TinyOS eases programming effort with a component library that contains several communication protocols such as IEEE 802.15.4, 6LoWPAN, and S-MAC, sensor drivers for supported platforms, and data storage.

Contiki is another event-based OS but it supports preemptive multi-threading through a library on top of the event-handler kernel. Contiki has a IPv6 protocol stack with 6LoWPAN header compression, IETF RPL IPv6 routing, and the IETF CoAP application layer protocol. The operating systems has a built-in MAC protocol referred to as ContikiMAC. Embedded applications are written against OS core functionality and can be dynamically loaded and unloaded at run time due to event-driven model and the use of Inter-Process Communication (IPC) messaging between components. Other supported features include Flash-based file system (Coffee) and software-based power profiling mechanism. A typical Contiki configuration requires 2 kB data and 40 kB program memory.

MANTIS is a preemptive OS that uses POSIX-like threads. Thus, it combines the familiar UNIX-style programming with energy-efficient scheduler suitable for duty-cycle sleeping of a sensor node. Its footprint is relatively small, requiring less than 500 B data and 14 kB program memory [1] .

5.2 Middlewares

Middleware is an application layer that consists of frameworks and interfaces that ease the application development, runtime configuration, and management in WSNs [25]. The sensor network middleware comprises the common functions and services found in sensor network applications such as data filtering, data analysis, localization, security, and collaboration between sensor nodes. Conceptually, middleware is located between application layer and a WSN OS.

A middleware may provide services for code, data, and resource management, service discovery, and storage. Code management handles the application code dissemination in the network. Data queries allow accessing gathered data. Resource management allows management of node peripherals and/or computational and/or energy resources. Service discovery service allows dynamic detection of resources and services. Storage services allow storing data into the sensor network.

Several types of middlewares have been proposed for WSNs. These are categorized in Figure 5.3. A *programming abstraction* middleware describes how the applications are created for a middleware. The abstraction can be global (network wide) or local (node specific). For example, a global middleware provides access to network resources, whereas a local middleware provides access to the node's internal resources, such as memory or sensors. Middleware *interfaces* aim to standardize hardware access, such as the transducer interface specification of IEEE 1451. As a special case, middlewares may use Markup Language (ML) to describe sensor applications and services with a high level, declarative language such as XML. This allows programming intelligent distributed sensor applications as the applications can understand also the context and meaning of sensor data.

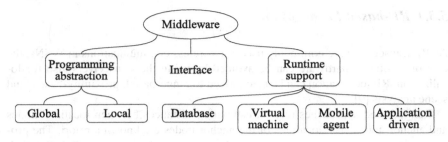

Fig. 5.3 Classification of WSN middleware.

Middlewares may also provide runtime database, virtual machine, mobile agent, or application drive services. *Database middlewares* use SQL-like interface to query data from the network. Thus, the network acts as a distributed database where nodes may partially fulfill the query. The approach is especially useful for complex, one-time queries. For example, TinyDB [18] is a query processing system implemented on top of the TinyOS. It supports basic SQL type query operations and data aggregation for improving network energy efficiency. *Virtual Machine (VM)* middlewares run applications as generic byte code, therefore allowing the same applications to run on heterogeneous hardware platforms e.g. with incompatible native executable formats. A *mobile agent* is a transferable object that in addition to the code carries its state and data. A mobile agent makes its migration and processing decisions autonomously. Typically, a mobile agent operates on top of a virtual machine to obtain platform independence and small object size. The *application driven* middlewares support task allocation, networking, and distributed computing. The component library of the TinyOS includes the application driven middleware functionality as it provides methods for network communication and distributed services, and abstracts data acquisition, allowing programmer to concentrate on implementing the sensing applications.

5.3 Localization

Ubiquitous localization has been widely studied during the recent years. In general, the solutions focus on finding effective location estimation algorithms and measurements that correlate with location. The underlying technologies vary from pure RF-based, to UltraSound (US), InfraRed (IR), and multimodal solutions.

5.3.1 RF-based Localization

As RF transceivers are commonly used for wireless communications in WSNs, the
cost of additional hardware can be avoided by using the transceivers also for localization. RF localization can be categorized to range-based, proximity-based, and
scene analysis [9].

Range-based approaches rely on estimating distances between localized nodes
and anchor nodes. The locations of the anchor nodes are known a priori. The process of localization using ranges is called lateration and is presented in Figure 5.4.
A common technique is to use Received Signal Strength (RSS) to estimate the
range between transmitting and receiving devices. Typically, all but the simplest and
the lowest cost WSN transceivers support RSS calculation. In the cases where the
transceiver does not support RSS, the signal strength can be estimated by sending
localization beacons with varying power levels [13]. As each beacon is marked with
the transmission power level, the beacon that was received with the lowest power
level packet denotes the RSS. Thus, the range-based approach can be easily applied.
As a drawback, the distance estimates using RSS can have large errors due to multipath signals and shadowing caused by obstructions [21]. This inherent unreliability
has to be addressed in the used localization algorithms.

Several location estimation techniques can be used in range-based localization.
Utilized methods include trilateration, weighted center of gravity calculation, and
Kalman filtering. Many mathematical optimization methods, such as the steepest
descent method, sum of errors minimization, and Minimum Mean Square Error
(MMSE) method, have been used to solve range-based location estimation prob-
lems.

Proximity-based approaches exploiting RF signals [12, 2] estimate locations
from connectivity information (Figure 5.5). Such solutions are also commonly referred to as range-free in the literature. In the strongest anchor method [12] (Figure

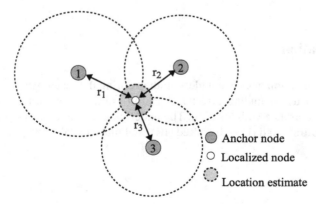

Fig. 5.4 In range-based localization the place is estimated in the intersection of measured distances
to the anchor nodes. This process is referred to as lateration.

5.5) the localized node is estimated to be located immediately near an anchor node in range. The method can be naturally combined with typical network protocols, as routing nodes act as anchors and localized nodes connect to one of the routers in range. Alternatively, the proximity-based approach may use multiple anchor nodes, e.g. to calculate location as geometric center [2]. The proximity-based localization is very simple to implement and, as ranging is not needed, hardware requirements are modest. However, as a drawback, the strongest anchor node method gives only a very coarse location, and the multi anchor solutions improve granularity only slightly. Thus, to reach small granularities the proximity-based schemes require a dense grid of anchor nodes.

Scene analysis consists of an off-line learning phase and an online localization phase as presented in Figure 5.6. The off-line phase includes recording RSS values corresponding to different anchor nodes as a function of the users location. The recorded RSS values and the known locations of the anchor nodes are used either to construct an RF-fingerprint database [17] or a probabilistic radio map [4, 32]. In the online phase the localized node measures RSS values to different anchor nodes. With RF-fingerprinting the location of the user is determined by finding the recorded reference fingerprint values in the signal space that are closest to the measured one. The unknown location is then estimated to be the one paired with the closest reference fingerprint or in the (weighted) centroid of k nearest reference fingerprints. Location estimation using a probabilistic radio map includes finding the point(s) in the map that maximize the location probability.

The applicability and scalability of scene analysis approach is greatly reduced by the time consuming collection and maintenance of the RF sample database. Searching through the sample database or radio map is computationally intensive. The Joint Clustering (JC) technique [32] uses location clustering to reduce the computational cost of searching the radio map. It improves the scalability of the searching algorithm to some extent. MoteTrack [17] achieves similar effect by disseminating the RF-fingerprint database to a WSN and decentralizing the localization procedure.

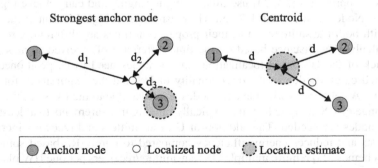

Fig. 5.5 Proximity-based localization. In the strongest anchor method, the location is at the closest anchor node. In the centroid method, the location is the geometric center of the anchor nodes in range.

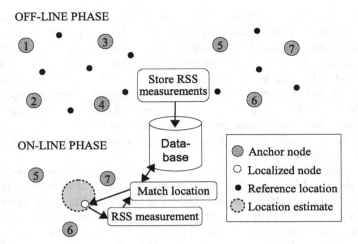

Fig. 5.6 Localization with scene analysis relies on pre-recorded, offline signal strength measurements that are matched online againsts node's measurements.

In general, RF signal strength based localization possesses fundamental limits due to the unreliability of the measurements [4]. There is strong evidence that, at best, accuracy in the scale of meters can be achieved regardless of the used method [4]. Time-of-flight can be used to improve localization accuracy to well below 1 m, although this requires accurate clocks. In practice, the time-of-flight ranging involves two-way message exchange between devices to calculate the elapsed time.

5.3.2 Localization with Ultrasound

US-based approaches [23, 24] use time-of-flight ranging and can achieve high accuracies. Nodes include both US and RF transmitters. A signal is sent at the same time with both transmitters but as their propagation times are different, a receiver may calculate the distance based on the time difference of received signals. As a drawback of the US-based localization, anchor nodes need to be positioned and orientated carefully due to the directionality of US and the requirement for LOS exposure. A dense network of anchor nodes is needed due to the LOS requirement, short range of US, and the fact, that typically ranging measurements to at least four anchor nodes are needed. The addition of US transmitters and receivers increases HW costs and reduces energy-efficiency compared to purely RF-based solutions. Some schemes [24] require multiple US transmitters/receivers per one HW platform further increasing the hardware costs.

5.3.3 Localization with Infrared

IR-based solutions [31] are based on inferring proximity. They can localize nodes inside the range of LoS IR transmissions. IR-based schemes suffer errors in the presence of obstructions. Also, differing light and ambient IR levels, caused by for example fluorescent lighting or direct sunlight, produce difficulties [23]. The anchor network costs are high because a dense matrix of IR sensors is needed in order to avoid dead spots.

5.3.4 Data Fusion

In the presence of a myriad of location sensing techniques data fusion has become and attractive location estimation method. It can combine measurements from multiple sensors while managing measurement uncertainty. For example, Fox et al. [6] survey Bayesian filtering techniques capable of multisensor fusion. Probabilistic fusion methods require relative large amounts of computation. Thus, in the presence of resource constrained nodes, a centralized implementation running in a more powerful base station is often the only feasible choice. For example in the Localization Stack [10] the fusion layer is implemented in Java.

5.4 Time Synchronization

Maintaining a coherent time base for each node in the network is essential for many sensing applications. This allows knowing the exact time instance of an occurred event, required e.g. in high precision sensing applications. Thus, the network time should be either real time or convertible to the real time e.g. at the gateway. Another use case for the synchronization is the comparison of occurred events. In that case, the network time does not necessary have to be real world time but rather expresses the order of events. This is required e.g. in object tracking.

The most straightforward way to achieve accurate synchronization would be to equip every node with a Global Positioning System (GPS) receiver, Universal Time Coordinated (UTC) signal receiver or an accurate atomic clock. However, in reality this would be infeasible in WSNs due to increased size, cost, and energy consumption. Thus, synchronization has to be achieved with a specific synchronization protocol.

Although effective in the Internet, the widely used Network Time Protocol (NTP) is too resource consuming for WSNs. Furthermore, it requires external configuration making ad hoc operation impossible. The IEEE 1588 standard for a precision clock synchronization protocol for networked measurement and control systems has the similar shortcomings.

In large networks, a synchronization protocol must act in multihop fashion. These multi-hop WSN time synchronization protocols can be categorized to tree-based, flooding-based, reference broadcasting, co-operative, delay-based time synchronization protocols. Synchronization protocols designed for WSNs and targeting at network-wide synchronization via multiple hops are discussed in the following.

5.4.1 Tree-Based Time Synchronization

Tree-based time synchronization protocols achieve global time using a tree structure to forward time information. The operating principle of these protocols is illustrated in Figure 5.7. In the tree-based synchronization, one node in the network acts as a time reference. Other nodes in the tree are synchronized to the time of the reference node by sending periodical messages through the tree structure. Parent nodes in the tree act as time sources for their child nodes in the next levels of the tree. The time information then progresses through the tree between adjacent levels in the tree.

Timing-sync Protocol for Sensor Networks (TPSN) [7] builds a hierarchical structure for achieving global time synchronization. The structure is constructed in a level discovery phase. Each node is assigned a level and the time reference node is at level 0. In the synchronization phase, the time reference node periodically initiates synchronization. Synchronization data is exchanged level by level. TPSN uses two messages to synchronize a pair of nodes in different levels. The disappearance of a time source is handled by re-transmitting synchronization requests a specific number of times and using level discovery to obtain new level. Reference node failure is taken into account by using a leader election algorithm when nodes at level 1 notice that the reference node is lost. The level discovery, the re-transmission during lost time sources, and reference node election increase messaging overhead and reduces energy-efficiency. This level assignment is fixed for the lifetime of the reference node. This reduces the robustness and makes the protocol unsuitable for applications with highly mobile nodes [8, 30].

Lightweight Tree-based Synchronization (LTS) [8] protocol achieves global time synchronization by creating a minimum height spanning tree. Minimum height spanning tree maximizes accuracy since every hop introduces error in the synchronization. Similarly to TPSN, a pair of nodes in the tree are synchronized with two messages. For time reference robustness multiple reference nodes are considered. However, this requires a leader election algorithm increasing communication overhead. Furthermore, the robustness is in direct correlation to the amount of backup reference nodes that need to be determined prior to deployment. Fault tolerance against dynamic channel variations, changes in topology, changes in size, and node mobility is achieved by re-creating the spanning tree on every re-synchronization. This increases messaging overhead significantly and reduces energy-efficiency. LTS provides also a de-centralized version where nodes can query time information from a reference node. In this, the messaging overhead is large since the queries have to

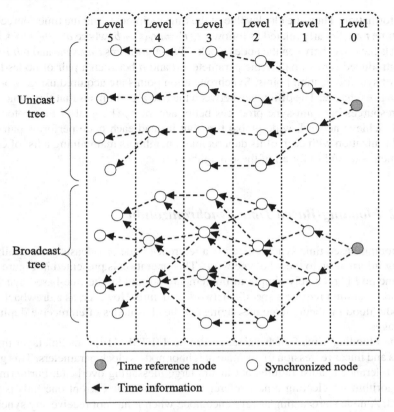

Fig. 5.7 Tree-based time synchronization protocols where nodes synchronize to the time of the root node.

be first forwarded to the reference node, possibly via multiple hops, and then synchronization is achieved by reversing the path of the query.

Delay Measurement Time Synchronization (DMTS) [22] uses one message to synchronize a sender and all the receivers in its neighborhood. For maximizing achievable accuracy the synchronization message is timestamped at low level just prior to transmission and right after reception. The multi-hop DMTS algorithm uses a leader selection algorithm to select a time reference for the whole network. The time reference is at time source level 0. The time reference periodically broadcast its time synchronizing its children at level 1. The children then broadcast their time synchronizing the nodes in the next level. This is continued until leaf nodes are reached. The algorithm inflicts unnecessary communication overhead to the time recipients since the messages are broadcast and also redundant messages from lower levels are received. Fault tolerance is not considered in the protocol.

The protocol proposed in [27] is based on the assumption that the clock value of a time recipient node can be converted to the clock value of a time source using linear transformation. Thus, the clock value of the time recipient t_r and the time source t_s at real-time t are linearly related as follows: $t_s(t) = at_r(t) + b$, where a is the clock drift and b is the offset between the nodes clocks. Two algorithms *tiny-sync* and *mini-sync* are introduced for estimating the parameters a and b between a pair of nodes from a set of synchronization points. Synchronization points are acquired using two-way message exchange. The difference between the two algorithms is that tiny-sync uses less resources but mini-sync produces better accuracy. The goal of the protocol is not to achieve network-wide global time but instead each node performs pairwise synchronization with each of its data recipient neighbors maintaining a list of clock parameters (a and b) for all of these neighbors.

5.4.2 Flooding-Based Time Synchronization

In flooding-based time synchronization a reference node broadcasts periodically its timing information to the whole network. This operation is presented in Figure 5.8. The method is relatively simple to implement and, unlike the tree-based synchronization, does not require a specific network structure. However, as a drawback, the periodic flooding incurs large messaging overhead as nodes often receive duplicate messages.

The Flooding Time Synchronization Protocol (FTSP) [19] uses link layer timestamps and linear regression to estimate neighbor node's clock parameters. This gives good tolerance against failures but incurs large messaging overhead. Furthermore, an algorithm for electing a new reference node when the current one fails is presented. A node starts acting as reference node when it has not receive any synchronization frames for a while. To ensure that a network has only one reference node, only the node with the smallest address remains active after the synchronization flooding resumes.

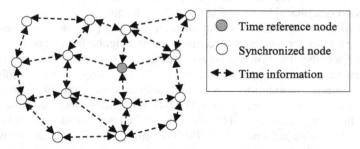

Fig. 5.8 Flooding-based time synchronization protocol broadcast synchronization messages to the whole network.

5.4.3 Reference Broadcasting

In the reference broadcasting, nodes are synchronized to a local reference node in their neighborhood (broadcast domain). In Reference Broadcast Synchronization (RBS) [5], the reference node broadcasts its time information to its immediate neighbors within one hop radius as shown in Fig. 5.9. To achieve global synchronization, RBS requires nodes that are synchronized to multiple broadcast domains. Thus, when forwarding message, these nodes convert timestamps between the different broadcast domains based on the differences between the clock parameters.

RBS removes several non-deterministic sources from traditional time synchronization, and is able to achieve high accuracy. However, the RBS method incurs messaging overhead similar to the flooding-based synchronization. The overhead is reduced with on-demand post-facto synchronization when continuous synchronization is not not needed. Still, when continuous synchronization is required, a constant messaging overhead occurs also in RBS.

TSync [3] is another reference broadcasting method that comprises a centralized Hierarchy Referencing Time Synchronization (HRTS) protocol, and de-centralized Individual-based Time Request (ITR) protocol. HRTS synchronizes network of nodes using one or multiple reference nodes. A time source synchronizes its neighbors with three message exchanges. A combination of receive-to-receiver synchronization similar to RBS and two-way message exchange similar to TPSN and LTS is used. Using a level indicator in the synchronization messages, nodes are synchronized from level to level in a multi-hop environment. While HRTS reduces the messages compare to multi-hop RBS, it still inflicts continuous messaging overhead. ITR is similar to de-centralized version of LTS and includes the same drawbacks.

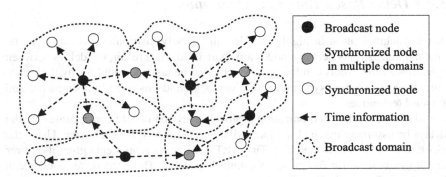

Fig. 5.9 Topology of multi-hop RBS. A neighborhood of nodes is synchronized using a broadcast node and exchanging between neighbors. Multi-hop synchronization is achieved by converting timestamps hop by hop and using nodes in multiple broadcast domains to forward data between the domains.

5.4.4 Co-operative Time Synchronization

Co-operative time synchronization does not use dedicated time reference nodes. Instead, the whole network is synchronized to an equilibrium time derived from the individual times of the nodes. This increases robustness as the network does not have a single point of failure. However, co-operative synchronization is usually complex, requiring a lot of messaging between nodes [30].

Time-Diffusion synchronization Protocol (TDP) [29] consists of active and inactive cycles. At the start of every active cycle a subset of nodes is selected as masters by an Election/Re-election Procedure (ERP). The timing messages sent by the masters create individual tree structures for every master. Furthermore, the protocol includes a method for detecting outliers. TDP does no rely on external time servers making it fully self-contained. Also, the creation of new synchronization trees in every cycle increases the fault tolerance. However, the messaging overhead inflicted by the protocol is significant and the convergence time high.

In Rate-based Diffusion Protocol (RDP) [16, 15], nodes achieve global synchronization by periodically spreading local synchronization information to the entire network. The protocol consists of synchronous and asynchronous variants. In the synchronous variant, a node exchanges its clock values with each neighboring node at one pass. The clocks are adjusted by a factor proportional to the time difference between every pair of neighboring nodes. In the asynchronous variant, an average of the clock values of a neighborhood of nodes is calculated and used as the clock value.

5.4.5 Delay-Based Time Synchronization

Time synchronization in highly mobile and sparsely distributed networks can be challenging, as links between nodes are short-lived. As a result, the delays between synchronization events can be long, which must be taken account for by the protocol. Delay-based time synchronization works around this issue by using store and forward techniques.

Ad hoc network synchronization protocol described in [26] compensates for the delays by assuming that nodes' clock drifts in relation to UTC are known. Using the clock drift value and Round Trip Time (RTT) delay measurements gives the lower and upper bounds for the message's multi-hop delay. Then, the delay bounds can be converted to the local time of the destination node. This operation principle is illustrated in Figure 5.10.

Meier et al. [20] propose an improved version of the ad hoc network synchronization protocol. In their work, a tighter lower bound for the delay is defined using an improved formula for the calculations. In addition, a method for defining optimal bounds is introduced. However, this methods assumes that message delay uncertainties are negligible and requires additional computation, communication and memory.

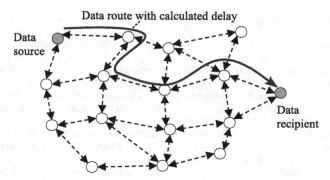

Fig. 5.10 Delay-based time synchronization. Event time is derived from the elapsed time in message delivery from a source to a target.

References

1. Bhatti, S., Carlson, J., Dai, H., Deng, J., Rose, J., Sheth, A., Shucker, B., Gruenwald, C., Torgerson, A., Han, R.: MANTIS OS: An embedded multithreaded operating system for wireless micro sensor platforms. In: ACM/Kluwer Mobile Networks & Applications (MONET), Special Issue on Wireless Sensor Networks (2005)
2. Bulusu, N., Heidemann, J., Estrin, D.: GPS-less low-cost outdoor localization for very small devices. Personal Communications, IEEE [see also IEEE Wireless Communications] **7**(5), 28–34 (2000)
3. Dai, H., Han, R.: Tsync: a lightweight bidirectional time synchronization service for wireless sensor networks. SIGMOBILE Mob. Comput. Commun. Rev. **8**(1), 125–139 (2004). DOI http://doi.acm.org/10.1145/980159.980173
4. Elnahrawy, E., Li, X., Martin, R.P.: The limits of localization using signal strength: a comparative study. In: Sensor and Ad Hoc Communications and Networks, 2004. IEEE SECON 2004. 2004 First Annual IEEE Communications Society Conference on, pp. 406–414 (2004)
5. Elson, J., Girod, L., Estrin, D.: Fine-grained network time synchronization using reference broadcasts. SIGOPS Oper. Syst. Rev. **36**(SI), 147–163 (2002). DOI http://doi.acm.org/10.1145/844128.844143
6. Fox, V., Hightower, J., Liao, L., Schulz, D., Borriello, G.: Bayesian filtering for location estimation. Pervasive Computing, IEEE **2**(3), 24–33 (July-Sept. 2003). DOI 10.1109/MPRV.2003.1228524
7. Ganeriwal, S., Kumar, R., Srivastava, M.B.: Timing-sync protocol for sensor networks. In: SenSys '03: Proceedings of the 1st international conference on Embedded networked sensor systems, pp. 138–149. ACM, New York, NY, USA (2003). DOI http://doi.acm.org/10.1145/958491.958508
8. Greunen, J.V., Rabaey, J.: Lightweight time synchronization for sensor networks. In: WSNA '03: Proceedings of the 2nd ACM international conference on Wireless sensor networks and applications, pp. 11–19. ACM, New York, NY, USA (2003). DOI http://doi.acm.org/10.1145/941350.941353
9. Hightower, J., Borriello, G.: Location systems for ubiquitous computing. Computer **34**(8), 57–66 (2001)
10. Hightower, J., Brumitt, B., Borriello, G.: The location stack: a layered model for location in ubiquitous computing. Mobile Computing Systems and Applications, 2002. Proceedings Fourth IEEE Workshop on pp. 22–28 (2002). DOI 10.1109/MCSA.2002.1017482
11. Hill, J., Szewczyk, R., Woo, A., Hollar, S., Culler, D., Pister, K.: System architecture directions for networked sensors. In: Proc. 9th ACM Int'l Conf. on Architectural Support for Program-

ming Languages and Operating Systems (ASPLOS'00), pp. 94–103. Cambridge, MA, USA (2000)

12. Hodes, T.D., Katz, R.H., Servan-Schreiber, E., Rowe, L.: Composable ad-hoc mobile services for universal interaction. In: MobiCom '97: Proceedings of the 3rd annual ACM/IEEE international conference on Mobile computing and networking, pp. 1–12. ACM, New York, NY, USA (1997). DOI http://doi.acm.org/10.1145/262116.262121

13. Kaseva, V.A., Kohvakka, M., Kuorilehto, M., Hännikäinen, M., Hämäläinen, T.D.: A wireless sensor network for RF-based indoor localization. EURASIP Journal on Advances in Signal Processing (2008). DOI 10.1155/2008/731835

14. Kuorilehto, M., Kohvakka, M., Suhonen, J., Hämäläinen, P., Hännikäinen, M., Hämäläinen, T.D.: Ultra-Low Energy Wireless Sensor Networks in Practice - Theory, Realization and Deployment. John Wiley & Sons Ltd (2007)

15. Li, M.Q., Rus, M.D.: Global clock synchronization in sensor networks. IEEE Trans. Comput. **55**(2), 214–226 (2006). DOI http://dx.doi.org/10.1109/TC.2006.25

16. Li, Q., Rus, D.: Global clock synchronization in sensor networks. In: INFOCOM 2004. Twenty-third AnnualJoint Conference of the IEEE Computer and Communications Societies, vol. 1, p. 564Ű574 (2004). DOI 10.1109/INFCOM.2004.1354528

17. Lorincz, K., Welsh, M.: MoteTrack: A robust, decentralized approach to RF-based location tracking. In: In Proceedings of the International Workshop on Location- and Context-Awareness (LoCA 2005) at Pervasive 2005. Oberpfaffenhofen, Germany (2005)

18. Madden, S., Franklin, M.J., Hellerstein, J.M., Hong, W.: The design of an acquisitional query processor for sensor networks. In: Proc. ACM Int'l Conf. on Management of Data (SIGMOD'03), pp. 491–502. San Diego, CA, USA (2003)

19. Maróti, M., Kusy, B., Simon, G., Ákos Lédeczi: The flooding time synchronization protocol. In: SenSys '04: Proceedings of the 2nd international conference on Embedded networked sensor systems, pp. 39–49. ACM, New York, NY, USA (2004). DOI http://doi.acm.org/10.1145/1031495.1031501

20. Meier, L., Blum, P., Thiele, L.: Internal synchronization of drift-constraint clocks in ad-hoc sensor networks. In: MobiHoc '04: Proceedings of the 5th ACM international symposium on Mobile ad hoc networking and computing, pp. 90–97. ACM Press, New York, NY, USA (2004). DOI http://doi.acm.org/10.1145/989459.989471

21. Patwari, N., Ash, J.N., Kyperountas, S., Hero III, A.O., Moses, R.L., Correal, N.S.: Locating the nodes: cooperative localization in wireless sensor networks. Signal Processing Magazine, IEEE **22**(4), 54–69 (2005)

22. Ping, S.: Delay measurement time synchronization for wireless sensor networks. Tech. Rep. IRB-TR-03-013, Intel Research Berkeley Lab (2003)

23. Priyantha, N.B., Chakraborty, A., Balakrishnan, H.: The cricket location-support system. In: MobiCom '00: Proceedings of the 6th annual international conference on Mobile computing and networking, pp. 32–43. ACM Press, New York, NY, USA (2000)

24. Priyantha, N.B., Miu, A.K.L., Balakrishnan, H., Teller, S.: The cricket compass for context-aware mobile applications. In: MobiCom '01: Proceedings of the 7th annual international conference on Mobile computing and networking, pp. 1–14. ACM Press, New York, NY, USA (2001)

25. Römer, K., Kasten, O., Mattern, F.: Middleware challenges for wireless sensor networks. ACM SIGMOBILE Mobile Computing and Communications Review **6**(4), 59–61 (2002)

26. Romer, K.: Time synchronization in ad hoc networks. In: MobiHoc '01: Proceedings of the 2nd ACM international symposium on Mobile ad hoc networking & computing, pp. 173–182. ACM, New York, NY, USA (2001)

27. Sichitiu, M., Veerarittiphan, C.: Simple, accurate time synchronization for wireless sensor networks. In: WCNC '03: Proceedings of the IEEE conference on Wireless Communications and Networking, vol. 2, pp. 1266–1273 (2003)

28. Stallings, W.: Operating Systems Internals and Design Principles, 5 edn. Prentice-Hall (2005)

29. Su, W., Akyildiz, I.F.: Time-diffusion synchronization protocol for wireless sensor networks. IEEE/ACM Trans. Netw. **13**(2), 384–397 (2005). DOI http://dx.doi.org/10.1109/TNET.2004.842228

30. Sundararaman, B., Buy, U., Kshemkalyani, A.D.: Clock synchronization for wireless sensor networks: a survey. Ad Hoc Networks **3**(3), 281–323 (2005). DOI DOI:10.1016/j.adhoc.2005. 01.002
31. Want, R., Hopper, A., Falcao, V., Gibbons, J.: The active badge location system. ACM Transactions on Information Systems **10**(1), 91–102 (1992)
32. Youssef, M.A., Agrawala, A., Shankar, A.U.: WLAN location determination via clustering and probability distributions. In: Pervasive Computing and Communications, 2003. (PerCom 2003). Proceedings of the First IEEE International Conference on, pp. 143–150 (2003)

Chapter 6
Sensor Data Collection

While some sensor networks can operate independently, e.g. by performing smart actuation based on measured sensor readings, in a typical use case sensor data is collected and utilized outside the WSN [16]. This necessitates data collection facilities and a connection between the sensor network and the rest of the world.

User interfaces, back-end services, and other applications located in the outside world connect to the sensor network via a gateway device as presented Figure 6.1. The gateway exposes an interface for accessing sensor networks. Thus, instead of communicating directly with a sensor node e.g. with Internet Protocol (IP), gateway may convert messages to a network specific protocol format. Instead of connecting directly to the gateway device, applications may access gateway through separate data collection services. These services provide additional functionality, such as database storage, that ease application development. In addition, the data collection services can provide an uniform interface for accessing WSNs instead of technology specific gateway interfaces. In practice, these services may locate on the same or different device as the gateway.

Typical sensor data collection services comprise:

- *Discovery of devices*: Identifying and enumerating the devices and sensors that are present in the network is the first step for data acquisition. In addition, the list of currently active devices can be used for diagnostics purposes.
- *Triggering data collection*: Data collection services allow determining what kind of measurements should be collected and how often. This avoids the collection of unnecessary data and its negative impact on network lifetime.

Fig. 6.1 Sensor networks are connected to the rest of the world via a gateway device. Applications utilizing the sensor data can connect directly to the gateway or use generic data collection services.

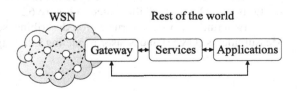

61

- *Sensor meta-data*: Understanding the meaning and significance of measured values requires meta-data on sensor's capabilities and properties. This meta-data includes sensor type, measured unit (e.g. Celsius or Fahrenheit), accuracy (e.g. $\pm 0.1^\circ$C), and other information relevant to the measurements.
- *Querying data*: Data queries can be performed in real-time or the requests may involve historical data. Thus, the data query services may utilize data archives (e.g. in database). Instead of raw values, the query services might use a high level language or advanced interfaces to refine the collected data, e.g. requesting the average temperature within last week would return only a single value.
- *Alerts*: As collected sensor values may indicate critical conditions that could cause loss of property, information, or lives e.g. when detecting a machine malfunction, intruder, or fire. Alerts provide a method to describe these situations and send a message to applications. For example, generates an alert when temperature rises over 50°C.

6.1 Gateway Interfaces

Constrained Application Protocol (CoAP) is an interface protocol especially targeted for constrained networks and machine-to-machine applications such as smart energy and building automation [12]. The protocol is designed to be simple and operates over User Datagram Protocol (UDP). CoAP provides request/response and subscribe/notify interaction models between application end-points. The protocol is designed according to the REST architecture. As such, the protocol is easy to map to Hypertext Transfer Protocol (HTTP) and URIs, therefore enabling web integrated sensors. Also, messages can be cached to improve performance, e.g. at the gateway to avoid requesting data directly from sensor network. Other features supported by CoAP are multicast support and congestion control.

6.2 Data Collection Services

6.2.1 Sensor Web Enablement

Open Geospatial Consortium (OGC) has developed a collection of sensor related eXtensible Markup Language (XML) encodings and interfaces referred to as Sensor Web Enablement (SWE) [3]. These specifications are summarized in 6.1 and their relations to each other are illustrated in Figure 6.2.

SWE defines two data formats for describing sensor data, Observations and Measurements (O&M) and Sensor Modeling Language (SensorML). O&M defines

Table 6.1 Encoding and interface specifications in OGC SWE.

Specification	Abbr.
Observations and Measurements	O&M
Sensor Modeling Language	SensorML
Transducer Markup Language	TML
Sensor Observation Service	SOS
Sensor Planning Service	SPS
Sensor Alert Service	SAS
Web Notification Service	WNS

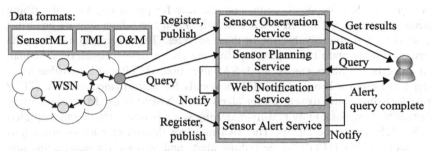

Fig. 6.2 Relations between OGC Sensor Web Enablement (SWE) specifications.

how observation[1] results are encoded. The encoding describes the observation procedure (the method of measurement), feature of interest (observation target), observed property (the phenomenon of observation), and the result. SensorML defines a model for describing devices and sensor capabilities. SensorML allows describing both physical and logical sensors (e.g. data filters, aggregators) by defining sensors and actuators as processes with certain input and output.

Sensor Observation Service (SOS) defines an interface for managing sensors and requesting observations from sensor networks. It has different profiles for accessing different kind of information: mandatory core profile for requesting meta-data and actual measurement data, optional transaction profile for registering a sensor to the system and inserting new measurement data, and optional enhanced operations profile for e.g. querying information regarding known observation or querying information of the observation target. SOS leverages the O&M model. From a user point of view, the sensor or procedure description is primarily metadata, which is only of interest to specialists during discovery, and then to assist evaluation or processing of individual results [29].

Sensor Planning Service (SPS) defines an interface for requesting user driven acquisitions and observations [33]. Compared to the SOS, it is to be used for more complex and potentially long lasting queries where results may not be returned immediately. The completion of the query can be detected either by polling the server

---

[1] In OGC terminology, a measurement is understood to be a special case of observation where the result is a numeric value.

or registering a notification to WNS. After the query is completed, the results can be acquired e.g. with SOS. SPS defines mandatory operation for a client to retrieve information about all parameters needed to submit the required query.

Sensor measurements in SWE are encoded with Transducer Markup Language (TML). A TML message contains both the actual data and the necessary information, such as the meta data of data provider, needed to understand it. It can be used for real-time streaming or carrying SOS or SPS results. For convenience, TML includes also additional transducer information such as characteristics (accuracy, coordinates, ...), calibration, operational conditions, and the relationships of transducers. Thus, TML can be used independently from O&M and SensorML.

Sensor Alert Service (SAS) defines an interface for publishing and subscribing to alerts from sensors [32]. In essence, it is a registry to which sensors advertise their services and clients subscribe to those services. The subscription defines the observed property and filters e.g. values that are above, below, equal, or not equal. All or none of the filters can be used. SAS uses both HTTP and Extensible Messaging and Presence Protocol (XMPP) protocols for performance reasons. HTTP is used for subscribing to sensors for alerts while XMPP is used for transporting the actual alerting data. Only real-time alerting is supported, meaning that an alert must be archived by the clients. The SAS interface defines mandatory operations (to be accessed via HTTP) for describing alerts and sensors, for subscribing and unsubscribing to alerts, and for renewing the subscription. Optional operations are defined for advertising the type of information to be published and retrieving WSDL definition of the server interface. Advertising can be cancelled or renewed.

Web Notification Service (WNS) delivers event notifications e.g. when a phenomena is observed [46]. Typically, the notifications comprise one-way message from a service to the registered clients. Additionally, WNS supports two-way communications where a service and a client may negotiate and exchange additional data upon an event. Notifications can be sent e.g. via e-mail, HTTP to other services, Short Message Service (SMS), fax, or Instant Message (IM).

6.2.2 Research proposals

OGC SWE has been implemented in Open Sensor Web Architecture (OSWA) [5], PULSENet [7], and WSN Application Service Platform (WASP) [8]. The common goal of these platforms is to ease the development of applications that utilize sensor data by defining a middleware layer for SWE. Additionally, WASP extends SWE with user authentication, WSN hardware configuration management, and application management for inserting, updating, and discovering sensor services.

SenseWeb [1] aims to provide an open infrastructure for registering and utilizing sensors. It uses a centralized architecture where sensors connect to a coordinator via gateways. The gateways hide network technology details and determine which data is available publicly. Compared to SWE, SensorWeb has less features. It supports only storing and querying sensor data, whereas alerts and other sensor related services are not specified. Furthermore, the platform does not support real-time streaming of sensor data making it mostly suitable for infrequent queries.

IrisNet [10] and Hourglass [13] are frameworks that allow complex queries over distributed Internet-connected sensors. Their goal is to reduce bandwidth usage to allow high bandwidth sensing, e.g video streams, with data filtering, smart query routing, and data caching. Both proposals implement smart query routing and data aggregation to optimize bandwidth usage. As a drawback, due to the dependency on IP and the requirement for significant computing power and storage in the routing devices, these frameworks are not suitable for all low power, resource constrained WSNs.

6.3 6LoWPAN

An alternative approach to the interaction between a WSN and the outside world is communicating directly with the sensor devices. In practice, this means using IP for end-to-end communications. The motivation for this approach is the use of sensor networks as a part of the global IP network. This allows the global scalability of WSNs and utilizing existing IP-based programs with the sensor networks.

Internet Engineering Task Force (IETF) has defined 6LoWPAN standard to describe how IP is used in IEEE 802.15.4 networks (Figure 6.3). It addresses IP routing on mesh networks and defines methods to allow transmitting large IP packets in bandwidth constrained environment. Other features addressed in 6LoWPAN include network autoconfiguration and multicast support.

The maximum frame size of IEEE 802.15.4 physical layer packet is 127 B. The usable portion of it is further reduced by IEEE 802.15.4 MAC layer overhead, leaving only B for upper layers. This causes several challenges for IPv6. First, as a IPv6 packet may be up to 1280 B, the packet may not fit into a single frame. To address this issue, 6LoWPAN defines fragmentation mechanisms for packets. Another challenge is the amount of overhead. IPv6 header is at least 40 B to which UDP used at the transport layer adds 8 B. Thus, without changes, the headers alone would consume a significant portion of available space. To reduce the overhead, 6LoWPAN defines header compression methods to be used with IPv6 and UDP headers. For example, 128 bit IPv6 addresses are compressed in 6LoWPAN by assuming that the 64 bit prefix part of the address is known by all nodes in the network and can thus be omitted. The remaining 64 bit are inferred from the 802.15.4 source and destination addresses. In a typical case, the 40 B IPv6 header can be compressed to 2 B[9].

IEEE 802.15.4 does not support multicast[2] functionality that is required for IPv6 compatibility. 6LoWPAN addresses this limitation by emulating the multicast with broadcasts.

---

[2] Multicast means the delivery of a message to a certain group of targets.

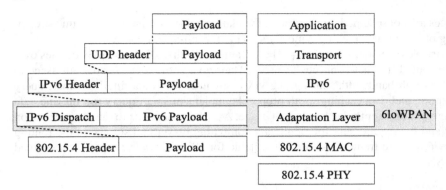

Fig. 6.3 6LoWPAN introduces an adaptation layer between IPv6 packets and IEEE 802.15.4 protocol.

6.4 Deployment Support and Diagnostics

Wireless networks suffer from limited communication capacity and unreliable communications e.g. due to wireless interference or long communication distances. Although some of the issues can be eliminated with a careful deployment, a practical network might also have software failures, logical errors in protocols and algorithms, and node failures due to energy depletion or hardware failures. Identifying problems in a large scale deployment is particularly challenging as problems may reflect to several parts of the network. For example, a broken link might cause congestion as the data is routed via an alternative route, therefore increasing traffic in the new route. This necessitates diagnostics to detect and identify the performance issues.

Different approaches to diagnose network reliability are presented in Fig. 6.4. The simplest method is to analyze whether data is received or not at the gateway. If the sensors are configured to collect periodic sensor data, the missing packets indicate problems. This kind of approach is technology independent but has limited usefulness as the reason for problems cannot be determined. Other approaches are passive monitoring with packet analyzers, the use of separate network for monitoring, and in-network diagnostics.

6.4.1 Passive monitoring

Passive monitoring tools are typically portable devices that listen to the WSN traffic. They are used both in protocol development and network deployments. The main benefits of the passive monitoring tools are that they do not cause any overhead to the monitored network and can be used without explicit planning. The simplest passive monitoring tools act as packet sniffers that decode and present packet exchange

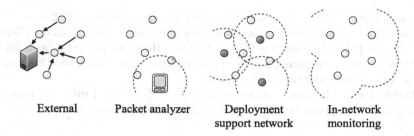

Fig. 6.4 Network operation can be diagnosed **a** externally from the collected sensor data, **b** locally with a packet analyzer, **c** with a separate deployment support network, or **d** with in-network diagnostics run on sensor nodes.

to the user. More advanced functionality includes bandwidth usage and connectivity analysis.

By combining data from several packet sniffers, it is possible to reconstruct network topology, determine bandwidth usage and routing paths, make connectivity analysis, and identify hot-spot nodes. For example, the approach proposed in [4] logs packets to Flash memory until the packet sniffers are manually retrieved later. Then, an offline analysis software merges and analyzes the packet traces. Other multi-sniffer approach proposed in [15] allows visualizing the otherwise complex behavior of a WSN with a graphical view. The proposed system is also able to replay the recorded network activities at different speeds.

6.4.2 Deployment Support Networks

Deployment Support Network (DSN) refers to a separate diagnostics network that is installed alongside the actual sensor network [2]. As such, the DSNs are typically short-lived and can be removed once a correct network operation is verified. A DSN collects diagnostics information by listening to the WSN traffic, but it may also inject traffic to the network e.g. to control the sensor nodes.

A DSN may be directly attached with wires to the WSN nodes. For example, the authors in [6] allow extensive debugging and controlling of WSN nodes. Their interactive debugging services include remote reprogramming, instant or timed Remote Procedure Call (RPC), and time-stamped data/event-logging. While the approach offers a versatile remote access to embedded node software, it does not capture WSN traffic.

Alternatively, a DSN may act as a network of passive monitoring devices. In that case, a DSN node has two radios: one for overhearing WSN traffic, and a second for forming the support network where the overheard packets are forwarded to a gateway.

LiveNet proposed in [4] can use wired Ethernet for collecting data. While this ensures that unreliability in wireless links do not affect diagnostics collection, it also increases the network deployment effort. Compared to the wired network, a wireless DSN is easier to deploy but may cause interference to the monitored WSN.

[11] and [6] define wireless DSNs that use Bluetooth scatternet with up to 100 mW transmission power. Due to the high performance radio, the DSNs have few weeks lifetime with two AA batteries.

Due to doubled hardware and increased costs DSNs are best suited for protocol testing and development but not for real-world long-term deployments.

6.4.3 In-Network Diagnostics

In-network diagnostics are built-in to the sensor nodes. That is, the nodes collect diagnostics information concerning their operation and pass it to the gateway using the same communication protocols and radios than the sensor data. The approach has two distinct benefits. First, additional diagnostics equipment is not needed which makes collecting the diagnostics throughout the lifetime of the network feasible. Second, nodes can include information about their internal operation and decisions making. The main drawback is that the diagnostics information consumes bandwidth in an already resource constrained network. Also, the diagnostics information cannot be forwarded when the node has connectivity problems and the diagnostics would be mostly needed.

The in-network diagnostics may be used to detect misconfigured or security compromised nodes, allow the remote debugging of sensor code, detect network performance or reliability problems, and detect anomalies in the sensor values. Generally, a good network diagnostics includes at least [14]:

- *Remaining lifetime*: The lifetime estimate of a node allows predicting when its power source has to be replaced, this way allowing proactive planning of service visits.
- *Neighbor information*: Listing node's neighbors within communication range and the link quality to each neighbors allow depicting the network topology and possible routing paths. In multihop topology, this information is essential in locating a possible network problem.
- *Traffic information*: that describes how much traffic is generated and forwarded at each node, thus allowing the detection of network congestion.
- *Network performance*: Performance diagnostics should include at least per link reliability information and forwarding latencies. These allow noticing reliability problems and assuring that the level of service meets the application demands.

As determining the level of performance and identifying potential network problems are important both for end-users and developers, many current network technologies support at least basic in-network diagnostics. As an example, ZigBee gateway specification [17] allows querying node's approximate energy level (near empty, half, full), and node's neighbors and link qualities.

Due to their tight reliability requirements, the common industrial WSN standards (e.g. WirelessHART, ISA100) have extensive diagnostics support. The supported features comprise an exact remaining lifetime estimate, and neighbor and traffic information. In addition, the performance information includes the average latency from gateway to device.

References

1. Aman Kansal Suman Nath, J.L., Zhao, F.: Senseweb: An infrastructure for shared sensing. IEEE Multimedia **14**(4), 8–13 (2007)
2. Beutel, J., Dyer, M., Meier, L., Thiele, L.: Scalable topology control for deployment-support networks. In: Fourth Int'l Symposium on Information Processing in Sensor Networks (IPSN 2005), pp. 359–363 (2005). DOI 10.1109/IPSN.2005.1440949
3. Botts, M., Percivall, G., Reed, C., Davidson, J.: OGC®sensor web enablement: Overview and high level architecture. In: S. Nittel, A. Labrinidis, A. Stefanidis (eds.) Second Int'l Conf., GSN 2006, Boston, MA, USA, October 1-3, 2006, Revised Selected and Invited Papers, pp. 175–190. Springer (2008)
4. Chen, B.r., Peterson, G., Mainland, G., Welsh, M.: Livenet: Using passive monitoring to reconstruct sensor network dynamics. In: S. Nikoletseas, B. Chlebus, D. Johnson, B. Krishnamachari (eds.) Distributed Computing in Sensor Systems, *Lecture Notes in Computer Science*, vol. 5067, pp. 79–98. Springer Berlin / Heidelberg (2008)
5. Chu, X., Buyya, R.: Service oriented sensor web. In: N.P. Mahalik (ed.) Sensor Networks and Configuration Fundamentals, Standards, Platforms, and Applications, chap. 3, pp. 51–74. Springer (2007). DOI 10.1007/3-540-37366-7
6. Dyer, M., Beutel, J., Kalt, T., Oehen, P., Thiele, L., Martin, K., Blum, P.: Deployment support network. In: K. Langendoen, T. Voigt (eds.) Wireless Sensor Networks, *Lecture Notes in Computer Science*, vol. 4373, pp. 195–211. Springer Berlin / Heidelberg (2007)
7. Fairgrieve, S.M., Makuch, J.A., Falke, S.R.: Pulsenet™: An implementation of sensor web standards. In: Int'l Symposium on Collaborative Technologies and Systems (CTS '09), pp. 64–75 (2009). DOI 10.1109/CTS.2009.5067463
8. Liu, J.C., Chuang, K.Y.: WASP: An innovative sensor observation service with web-/GIS-based architecture. In: 17th Int'l Conf. on Geoinformatics, pp. 1–6 (2009). DOI 10.1109/GEOINFORMATICS.2009.5293493
9. Montenegro, G., Kushalnagar, N., Hui, J., Culler, D.: Transmission of IPv6 packets over IEEE 802.15.4 networks. RFC 4944 (2007)
10. Nath, S., Deshpande, A., Ke, Y., Gibbons, P.B., Karp, B., Seshan, S.: Irisnet: an architecture for internet-scale sensing services. In: VLDB '2003: Proceedings of the 29th international conference on Very large data bases, pp. 1137–1140. VLDB Endowment (2003)
11. Ringwald, M., Römer, K., Vitaletti, A.: Passive inspection of sensor networks. In: J. Aspnes, C. Scheideler, A. Arora, S. Madden (eds.) Distributed Computing in Sensor Systems, *Lecture Notes in Computer Science*, vol. 4549, pp. 205–222. Springer Berlin / Heidelberg (2007)
12. Shelby, Z., Hartke, K., Bormann, C., Frank, B.: Constrained application protocol (coap). IETF RFC draft, draft-ietf-core-coap-07 (2011)
13. Shneidman, J., Pietzuch, P., Ledlie, J., Roussopoulos, M., Seltzer, M., Welsh, M.: Hourglass: An infrastructure for connecting sensor networks and applications. Tech. rep., Harvard University (2004). Harvard Technical Report TR-21-04
14. Suhonen, J., Hänninen, M., Hämäläinen, T.D., Hännikäinen, M.: Remote diagnostics and performance analysis for a wireless sensor network. In: IEEE Workshop on Signal Processing Systems (SiPS), p. 6 (2011)

15. Yang, Y., Xia, P., Huang, L., Zhou, Q., Xu, Y., Li, X.: Snamp: A multi-sniffer and multi-view visualization platform for wireless sensor networks. In: Industrial Electronics and Applications, 2006 1ST IEEE Conference on, pp. 1–4 (2006). DOI 10.1109/ICIEA.2006.257222
16. Yick, J., Mukherjee, B., Ghosal, D.: Wireless sensor network survey. Computer Networks (52), 2292–2330 (2008)
17. ZigBee Standards Organization: Network Device: Gateway Specification (2011). ZigBee Document 075468r35

Chapter 7
Experiments

This section presents TUTWSN as a case study of WSN performance. TUTWSN uses cross-layer design between network protocols and hardware platforms to achieve the requirements of target WSN applications.

Two variations of the protocol are presented: low-energy and low-latency TUT-WSN. The design requirements for these variants are presented in Table 7.1. The low-energy TUTWSN is targeted at sensing applications requiring moderate throughput, long network lifetime, and forwarding latencies that are in the order of seconds per hop[2, 5]. The low-latency TUTWSN is targeted at localization and target tracking applications requiring very low end-to-end delays and light throughput[1]. It uses a heterogeneous approach where ultra low power mobile nodes are used with more energy consuming router nodes that form a backbone network. Both protocol variants support multihop networking with one or more sinks and share a common hardware platform. The program memory usage of the low-energy and low-latency TUTWSN protocols were 100 kB and 60 kB, respectively. The data memory usage was less than 4 kB in both cases. Thus, the protocols represent the feasibility of the resource constrained WSN hardware.

The platform used in the experiments is built from COTS components. A TUT-WSN node is controlled by a 8-bit Microchip PIC18LF8722 MCU and Nordic Semiconductor nRF24L01 transceiver. The transceiver is used with 1 Mbit/s data rate, which ensures short transmission and reception times that allow a node to spend most of its time in low-power states. A node is powered by two AA batteries.

7.1 Low-Energy TUTWSN

The low-energy TUTWSN uses a clustered topology, in which each cluster operates on its own frequency referred to as a cluster channel. This increases scalability and avoids collisions between clusters. In the experiments, TUTWSN MAC utilized 2 s access cycle, 4 ALOHA slots, and 8 reserved slots. Frame size was 32 B due to

Table 7.1 Design requirements of TUTWSN low-energy and low-latency protocols.

TUTWSN variant	Network size	Measurement interval	End-to-end latency	Router power	Node power	Lifetime
Low-energy	Thousands	30 s-15 min	<10 min	Battery	Battery	2 years
Low-latency	Hundreds	0.5 s-10 s	<6 s	Mains	Battery	4 years

hardware limitations. As data is sent only in the reserved slots, the total throughput at the MAC layer was 1 kbit/s.

The TUTWSN routing protocol uses the cost-based approach. The cost-based operation principle is extended by supporting several sinks as a separate cost is maintained for each sink. A node initially searches its neighbors with a network scan. When a new neighbor is found, a node requests the list of its known sinks and routing costs. This way, a node can change its next hop towards a sink when a lower cost is found. Additionally, nodes periodically recalculate the cost and broadcast an advertisement to their neighbors. As a result, nodes can react to the changes in the network conditions as cost changes.

7.1.1 Scalability

In a low-duty cycle MAC protocol (such as IEEE 802.15.4 or TUTWSN MAC), the maximum number of nodes (α) in an interference area can be determined by the access cycle length (T_{AC}), the superframe length, the average number of member nodes in each cluster, and the number of utilized non-interfering frequency channels (n_{CH}) as

$$\alpha = \frac{T_{AC} n_{CH} (1 + n_S)}{t_{SF} + t_{guard}}, \qquad (7.1)$$

where t_{SF} is the length of a superframe, t_{guard} is a short guard time between consecutive superframes. α is maximized by minimizing the superframe and guard time lengths and by maximizing T_{AC}, n_{CH}, and n_S. It can be clearly seen in the equation that by utilizing a high data-rate radio operating at a wide frequency band provides the highest scalability.

In the experimented 2.4 GHz low-energy TUTWSN, $T_{AC} = 4$ s, $t_{SF} = 280$ ms, and each superframe is followed by a 220 ms long guard period to allow time for data processing. In TUTWSN sensor node platforms, the interference range in indoor conditions is around 100 m equaling to the area of $31400 \, m^2$. 8 member nodes can be connected to each cluster head. The limit is due to data memory as various statistics is kept from each member. The transceiver provides 82 channels with 1 MHz channel separation. In practice, 41 of these channels can be considered as non-overlapping (2 MHz channel separation). Thus, 2880 nodes can be located within an interference area. If only one channel were used ($n_{CH} = 1$), α would be reduced to 72 nodes per an interference area.

The network depth is limited by the routing protocol. As routing cost is expressed in 8-bit integer value and per-hop cost ranges between 1 ... 8, the maximum network depth is 32.

7.1.2 Power Consumption

The power consumption of the platform was tested in a multihop network consisting of 2 sinks, 13 headnodes, and 12 subnodes (27 nodes in total). Each node measured its temperature and transmitted a packet every 10 s to the nearest sink. The maximum hop count was 3.

Average headnode and subnode power consumptions with 2 s, 4 s, and 8 s access cycle lengths are presented in Table 7.2. A short access cycle consumes more power as beacons must be sent and received more frequently. Depending on the used access cycle length, the lifetime with a conservative estimate of 2000 mAh battery capacity is estimated between 4.5 and 6.6 years for a subnode and 0.9 and 2.1 years for a headnode. The power consumption of a headnode is significantly higher as it must also receive channel during CAP and forward traffic.

7.1.3 Availability and End-to-end Reliability

The practical performance of the low-energy network was measured in an indoor deployment consisting of 120 nodes and 8 sinks. Nodes sent data in 60 s intervals and diagnostics data (e.g. battery voltage, buffer usage, and link reliabilities) in 120 s intervals. As a result, a node generated data packet on average in 40 s intervals. Each node transmitted its data to the nearest sink. An average routing path length was 4 and the maximum hop count to the sink was 6.

The reliability of the network was estimated with availability metric[4]. If a node does not have problems, the reception interval at a sink equals to the data generation interval. Retransmissions and packet drops increase the interval to reach a certain availability.

The average availability of nodes, and the availability of the most and least reliable nodes are presented in Fig. 7.1. On average, 95% of traffic is received in less than 140 s interval. The availability increases only slightly after 99% as the reception interval is increased, denoting that 1% of traffic is lost. The unreliability is caused by the limited data memory, each node can buffer only 20 packets. Therefore, buffer overflows cause packet drops, especially when a node has lost a next hop link but is still receiving traffic from other nodes.

Table 7.2 Average power consumption and estimated lifetime with a 2000 mAh battery.

| Access cycle length | Power consumption (μW) | | Lifetime (years) | |
	Subnode	Headnode	Subnode	Headnode
2 s	153	740	4.5	0.9
4 s	120	533	5.7	1.3
8 s	103	327	6.6	2.1

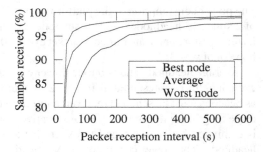

Fig. 7.1 Node availability in the experimented network.

7.2 Low-Latency TUTWSN

The low-latency network consists of router nodes and mobile nodes as shown in Fig. 7.2. The router nodes are responsible of data forwarding via a multihop network to one or multiple sinks. The low-latency TUTWSN operates on single network wide communication channel. The channel access method of routers and mobile nodes are different. The communication between routers is realized with a slotted ALOHA[3] protocol. The ALOHA protocol was selected to allow implementation with low-cost and low-power radio transceivers that do not support carrier sensing. When not transmitting data, a router listens to the channel continuously to allow fast data forwarding for mobile nodes.

The channel access of a mobile node is designed to minimize energy consuming idle listening. A mobile node broadcasts beacon frames periodically but with slightly randomized intervals to avoid collisions. After the beacon transmission, a node briefly listens to the channel for downlink data. The beacon can be piggy-

Fig. 7.2 Low-latency TUT-WSN topology. The router nodes forward data via multiple hops to one or multiple sinks. Mobile nodes broadcast their data to the router nodes.

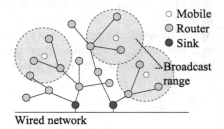

backed with application generated data. A router node that receives a piggybacked beacon waits a random time to avoid collisions and sends an acknowledgment to the mobile node. If the beacon frame containing data is not acknowledged, a mobile node temporarily shortens its beacon generation interval, thus reducing the delay between retransmission attempts.

Tge low-latency TUTWSN uses cost-based routing protocol. The routing and MAC protocols are cross-layer designed to lower delays and improve implementation efficiency. The routing layer uses network beacons to communicate cost information to neighbors and acquire in depth information from the neighborhood. The protocol layers share the routing table thus allowing fast lookup of next hop without without the need to cycle the packets between layers. This reduces queuing, processing, and stack handover delays.

7.2.1 Power Consumption and Network Lifetime

As the routers are active all the time, their power consumption does not depend on network behavior nor operating mode. The average measured power consumption for a router node is 78 mW. With the conservative estimate of 2000 mAh battery capacity the lifetime of a router is estimated to 8 days. Thus, in practice, the routers should be equipped with big enough batteries or be mains powered.

The average mobile node power consumption was measured using 0.25 s, 1 s, 4 s, 16 s, and 32 s access cycle lengths. The results are presented in Fig. 7.3(a). The lifetimes of a mobile node with the same access cycles and 2000 mAh battery capacity are shown Fig. 7.3(b). The lifetime ranges from 3.1 months to 2.1 years. As a node might not need to transmit every 0.25 s for its whole lifetime, a longer lifetime could be achieved by using shorter access cycles only when needed.

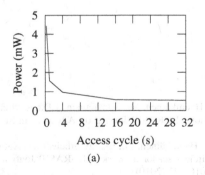
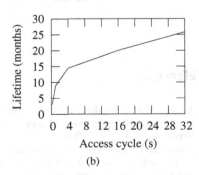

Fig. 7.3 Estimated mobile node power consumption and lifetime with 2000 mAh batteries.

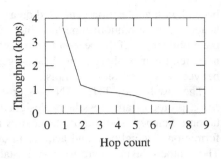

Fig. 7.4 Fixed multi-hop
scenario throughput results.
Packets were relayed over
1-8 hops. The data always
originated from the end of the
hop chain

7.2.2 Delay and Throughput

The performance of the low-latency network was tested using a network where the
nodes communicated via a fixed multi-hop route to a sink. This enables an accurate
analysis of the network behavior over varying amount of hops.

The delay results are presented in Table 7.3. The scenario consisted of an eight-
hop chain of nodes. 99% of the packets reached the sink in 2.1 s and all the packets
were received within 4.4 s.

Fig. 7.4 presents the throughput as a function of hop count. In the experiments,
the data always originated from the end of the hop chain. The throughput ranges
from 3.5 kpbs to 500 bps.

Table 7.3 Fixed multi-hop scenario delay results. Packets were relayed over 1-8 hops.

Percentile	Maximum delay (s)	Average delay per hop (s)
99%	2.1	0.4
99.5%	2.5	0.4
99.9%	3.6	0.5
100%	4.4	0.7

References

1. Kaseva, V.A., Kohvakka, M., Kuorilehto, M., Hännikäinen, M., Hämäläinen, T.D.: A wireless
   sensor network for RF-based indoor localization. EURASIP Journal on Advances in Signal
   Processing (2008), 28 (2008)
2. Kohvakka, M., Suhonen, J., Hämäläinen, T.D., Hännikäinen, M.: Energy-efficient reservation-
   based medium access control protocol for wireless sensor networks. EURASIP Journal on
   Wireless Communications and Networking (2010), 22 (2010)
3. Roberts, L.: ALOHA packet system with and without slots and capture. ACM SIGCOMM
   Computer Communication Review 5(2), 28–42 (1975)
4. Suhonen, J., Hämäläinen, T.D., Hännikäinen, M.: Availability and end-to-end reliability in low
   duty cycle multihop wireless sensor networks. Sensors 9(3), 2088–2116 (2009)

5. Suhonen, J., Kohvakka, M., Hännikäinen, M., Hämäläinen, T.D.: Design, implementation, and experiments on outdoor deployment of wireless sensor network for environmental monitoring. In: Proc. Embedded Computer Systems: Architectures, Modeling, and Simulation (SAMOS VI), *Lecture Notes in Computer Science*, vol. 4017, pp. 109–121. Springer (2006)

Chapter 8
Summary

The requirements of a WSN are diversified. The magnitude of data collected from a sensor network vary from bytes per day (e.g. environmental sensing) to kilobits per second (e.g. camera image streaming in surveillance). Further, the number of nodes in a network varies from few nodes to thousands of nodes. In addition, as the Quality of Service (QoS) requirements, such as latency and reliability, depend on application, there is no general purpose WSN technology that fits all use cases.

Although several platforms, standards, and proprietary technologies have been proposed, WSNs do not have a de-facto technology. While some of the techniques compete with each other, e.g. ZigBee and Z-Wave standards in home automation, most of them are targeted at specific use cases. Thus, platform components and communication protocols must be specifically selected to meet the application demands.

In low-power platforms, transceiver consumes most of the energy. Thus, it is possible to lower energy consumption by trading communication time with processing time with data preprocessing, data fusion, and aggregation. Accelerating computing intensive tasks, such as encryption, would leave processing power for other task. In addition, accelerated compression would allow more complex sensing, as images, sound, and video could be transferred energy-efficiently. ASIC implementation of communication protocols and applications could improve performance and decrease power consumption. However, as a trade-off, the network would lose its reprogrammability capabilities and configurability. Thus, in many cases, the use of COTS components is the most suitable solution.

The limitations in the current manufacturing techniques cause a trade-off between size, price, performance, and lifetime of a sensor node. An ideal WSN platform would combine the hundreds of MIPS processing power and several MBs memory of the high performance platforms with the 1 mA average and 1 μA sleep mode currents of low-energy platforms. However, it is unlikely that this can be achieved in the near future only by improving manufacturing technologies. This necessitates the use of energy-efficient, resource constrained protocols. With suitable protocols and currently available COTS components, a multihop mesh WSN can achieve the lifetime of several years with AAA batteries.

Overall, the processing power and capacity of a single sensor node is small. However, the large number of sensor nodes enable massively parallel distributed data processing and storage. Thus, the feasibility and useful features of WSNs lie in the co-operation of the sensor nodes.

Index